Pavement Asset Management

Scrivener Publishing
100 Cummings Center, Suite 541J
Beverly, MA 01915-6106

Publishers at Scrivener
Martin Scrivener(martin@scrivenerpublishing.com)
Phillip Carmical (pcarmical@scrivenerpublishing.com)

Pavement Asset Management

Ralph Haas and W. Ronald Hudson
with Lynne Cowe Falls

Scrivener
Publishing

WILEY

Co-published by John Wiley & Sons, Inc. Hoboken, New Jersey, and Scrivener Publishing LLC, Salem, Massachusetts.
Published simultaneously in Canada.

For general information on our other products and services or for technical support, please contact our Customer Care Department within the United States at (800) 762-2974, outside the United States at (317) 572-3993 or fax (317) 572-4002.

Wiley also publishes its books in a variety of electronic formats. Some content that appears in print may not be available in electronic formats. For more information about Wiley products, visit our web site at www.wiley.com.

For more information about Scrivener products please visit www.scrivenerpublishing.com.

Cover design by Kris Hackerott

Library of Congress Cataloging-in-Publication Data:

ISBN 978-1-119-03870-2

Printed in the United States of America

10 9 8 7 6 5 4 3 2 1

Dedication

This book is dedicated to the many practitioners, educators and researchers who have made a difference in advancing pavement management over the past five decades. We name a few in the following, with apologies to other deserving individuals whom we have unintentionally missed, and with recognition of the many planning, design, materials, construction and maintenance people who have contributed in various ways but could not all be realistically listed.

Fred Finn, Consultant; Roger Leclerc, Washington State DOT; Paul Irick, TRB; Bill Carey, TRB; Frank Botelho, FHWA; Frank McCullough, UT Austin; Roger Smith, Texas A & M; Katie Zimmerman, ApTech Consultants; Sue McNeil, U Delaware; Charlie Duggan, Connecticut DOT; Mo Shahin, Corps of Engineers; Harold Von Quintus, Applied Research Associates, Inc.; Stuart Hudson, Agile Assets Consultants; Gerardo Flintsch, Virginia Tech; Oscar Lyons, Arizona DOT; Dale Petersen, Utah DOT; Waheed Uddin, U Mississippi; Gary Elkins, AMEC Environment & Infrastructure; Harvey Treybig, ARE Consultants; George Way, Arizona DOT; Eric Perrone, Agile Assets Consultants; Mike Darter, U Illinois; Charles Pilson, Agile Assets Consultants; Bob Lytton, Texas A & M; Dave Luhr, Washington State DOT; Joe Mahoney, U Washington; Judith Corley Lay, N Carolina DOT; Linda Pierce, ApTech Consultants; Billy Connor, Alaska DOT

And

Bill Phang, Ontario DOT; Frank Meyer, Stantec Consultants; Matt Karan, Stantec Consultants; Bill Paterson, World Bank; Alex Visser, U Pretoria; Hernan de Solminihac; U Catholica, Santiago, Chili; Bert Wilkins, British Columbia DOT; Robert Tessier, Quebec Ministry of Transport; Tom Kazmierowski, Ontario DOT; Bruce Hutchinson, U Waterloo; Theuns Henning, U Auckland; Susan Tighe, U Waterloo; John Yeaman, Consultant Australia; Pim Visser, Consultant Netherlands; Cesar Queiroz, World Bank; Tien Fwa, U Singapore; Atsushi Kasahara, U Hokkaido; Martin Snaith, U Birmingham; Henry Kerali, U Birmingham; Rick Deighton, Deighton Consultants; Donaldson MacLeod, Public Works Canada.

RH, WRH and LCF

Contents

viii Contents

Part Seven: Looking Ahead

Preface

Pavement Management Systems by Haas and Hudson (1978) laid a foundation for using the systems methodology in a pavement management context. *Modern Pavement Management* by Haas, Hudson, and Zaniewski (1994)[1] built on the concepts of the original book but was a complete update of the original book. While there have been many advances in pavement engineering and management concepts since 1994, the basic structure of the pavement management process is largely intact. Therefore, the purpose of this book on *Pavement Asset Management* is to reflect current pavement engineering and management concepts and practice.

Although the concept of applying the systems method to pavement engineering and management has existed for several decades, there is still a need to make the case for adopting pavement management systems. Subsequent years saw pavement management systems broadly accepted and implemented by agencies and organizations with responsibilities for designing, constructing, and maintaining pavement structures. In fact the management systems concept has been and continues to be broadly implemented to the entire transportation and indeed civil infrastructure, as described in *Public Infrastructure Asset Management* by Uddin, Hudson and Haas (2013).[2]

Initial pavement management systems focused on the pavement design problem, i.e. what is the "best" pavement solution for a specific section of road. However, it was soon recognized that the systems method could be applied for selecting and programming what, where, and when projects should be selected for the optimum allocation of funds to a network

---

1 Haas, R., W.R. Hudson and J.P. Zaniewski, *Modern Pavement Management, Krieger Press, Florida, 1994.*

2 Uddin, W., W.R. Hudson, R.C.G. Haas, *Public Infrastructure Asset Management,* Second Edition, McGraw Hill Education Publications, New York, 2013.

of pavements managed by an agency. The two applications of the systems method to pavement management were termed "project" and "network" level pavement management. Subsequently, the capability of within-project alternatives was added to recognize that some network level systems were capable of identifying optimal levels of resources over time and between the different pavement strategies, but it did not have a mechanism for the actual selection of the timing and location of specific treatments. The confluence of pavement engineering at the project level and the management problem at the network level results in what may best be termed as good engineering-management.

To some extent, the separation of pavement design and management into discrete elements was an artifact of the technology available in the 1980s and 1990s. Specifically, the data and analysis methods needed for a project level design system were too complex, computer intensive, and time consuming for application at the network level. With the evolution of technology, the pavement design and engineering-management system process may be viewed as a continuum that ranges from the greatest level of data detail needed for a research project to the greatest level of aggregation, which is suitable for programming decisions at the national level.

Extension of the continuum concept in the pavement design and engineering-management process is complex and difficult to fully understand by any individual; hence, engineers and managers face the conundrum of selecting the content and level of detail needed in a text about pavement management systems. For example, there is no intention to make this a pavement design textbook. On the other hand, knowledge of pavement design is necessary for understanding the broader pavement engineering-management process at both the project and network levels.

In many areas of the overall pavement engineering-management process, we have made arbitrary decisions as to the level of detail presented in both the original books of 1978 and 1994, and in this book. This is necessary as the subject is too extensive to be fully treated in one book. The intention is to provide a holistic treatment of the process, with sufficient information on the various related topics for understanding and using the PMS process.

When the original books were published in 1978 and 1994, there was a need for a comprehensive document about pavement management systems. Relatively limited resources were available to engineers, managers, and educators about pavement management. Few organizations were actively pursuing and implementing pavement management systems at that time. To expand knowledge, the Federal Highway Administration sponsored a pavement management workshop for state highway agencies in Phoenix, Arizona, and Charlotte, North Carolina, in 1981. But in general, pavement management was not widely understood and embraced by administrators of highway agencies, the pavement engineering community, and academicians. In the intervening years there has been a plethora of publications about pavement management. There is now so much information (some good, some erroneous) about pavement management systems that it is difficult for a student or professional to know where to start and how to approach understanding, development, and use of pavement engineering-management systems.

This book is intended to present relevant current and new information needed for studying and applying pavement management systems.

Many people have contributed to this book. We have attempted to recognize as many as possible but will undoubtedly miss some, for which we apologize. Likewise, we have tried to condense or summarize some of the material as much as possible. Any resulting errors are the sole responsibility of the authors and not the contributors.

Recognition and special thanks are due to Dr. John Zaniewski who contributed in the early stages of this book including the outline and Sections of Part Two, but John was unable to join us as a co-author.

Special thanks are also due to Jan Zeybel and Shelley Bacik for their diligent and patient work on the many drafts of our manuscript. Thanks also to our Editor, Hank Zeybel, for his strong editorial work to produce a copy edited final draft, and to our publisher Phil Carmical of Scrivener with whom we were fortunate to be able to work in Austin, Texas.

Technical material has been contributed by Roger Smith and his team, and by Alan Kercher, Katie Zimmerman, Steve Seeds, Maggie Covault, Mike McNerney, Charles Pilson, Eric Perrone, Stuart Hudson and his team. Their contributions are very much appreciated.

Thanks are also due to the many hundreds of persons who have contributed to the advancement of PMS through development, use, implementation, and research over the last half century. Many are referenced in the book. We regret the inadvertent admission of any others.

Ralph Haas   W. Ronald Hudson   Lynne Cowe Falls

Part One

THE EVOLUTION OF PAVEMENT MANAGEMENT

1

Introduction

Many advances in the planning, design, construction, and maintenance of pavements have occurred in the past century. Pavement management, as practiced today as part of overall asset management, has evolved from early rudimentary efforts in the 1960s to a comprehensive technology, economic, and business-based process.

The first two books on PMS were published in the 1970s [1,2] and in many ways were a catalyst for ensuring developments and implementation of pavement management systems worldwide. Related documents include many guides, manuals, reports, and a vast array of publications, most of which can be accessed on agency websites.

Quite recently, the Canadian Pavement Asset design and Management Guide [3] has provided a valuable tool for practitioners and for college and university level instruction.

The last major PMS book, *Modern Pavement Management*, published in 1994, is comprehensive in scope and content and is still used in both university and professional environments [4]. In universities it is used as a text for senior and graduate level classes. Professionals use it to study the broad concept of pavement management systems, either by self-study or in a workshop environment.

Since 1994, there has been a transition in application of pavement management systems. Large agencies at the national and state level continue to use pavement management systems as a vital part of their asset management strategy in fulfilling their responsibility to society. This practice has also been transmitted to local and city agencies with pavement and other assets responsibility.

However, application of PMS in all areas of the public sector has migrated from project-level PMS to broader application at the network level.

As a result of this transition, it seemed clear to the authors that this book should deal primarily with the network-level PMS and so it does. Since the basic concepts and approach from 1994 still apply, this book picks up changes, improvements, and application developed since 1994. As a comparison, [4] provides the content for basic PMS studies, while this book updates concepts and applications for advanced studies.

In other words, the authors do not repeat the basic pavement design models and concepts. The reader may obtain those in [4,5]. The design models covered herein relate to MEPDG [6].

This book explains the development of asset management as it stemmed from pavement management in Chapter 46 of [4] but it does not cover asset management details that are presented in a book by [7].

2

Birth and Teen Years
of Pavement Management
(1967–1987)

Pavement management was born in the mid 1960s largely in response to numerous unanticipated pavement failures on the US Interstate and Canadian Highway Systems. These roads had been designed and constructed using the best known pavement design technology at that time, including the results of the $30 million AASHO Road Test. After an intensive national review of problems observed, the impossibility of making accurate single-point predictions of pavement performance due to national statistical variability of the major inputs became clear. Design methods at that time required as inputs estimated traffic, projected as-constructed materials properties, and estimated environmental conditions for a 20–30 year life of the pavement. These methods did not take into account the effects on performance of pavement maintenance, nor did they consider the life-cycle cost past the initial design period to include one or more overlays and rehabilitation activities, which everyone knew were common practice on heavy duty pavements. In response to this problem the National Cooperative Highway Research Program (NCHRP) funded a major research project (AASHTO) to find the reason and a solution for the problem [8].

A number of prominent civil engineers were involved in the US space program in the early 1960s and in various brainstorming sessions of the pavement "design" and "performance" problem. After many hours of discussion, they recognized the need to integrate planning, design, construction, maintenance, and rehabilitation into a coordinated systematic method for providing the required pavement performance over a 30, 40 or 50 year life. Figure 2.1 was the first input/output diagram developed to describe what has become known as a Pavement Management System (PMS). It illustrates the many important factors that govern pavement performance. A detailed description of the diagram is beyond the extent of this paper [2,4,9].

In a parallel study in Canada, what was then the Roads and Transportation Association of Canada also recognized the concept and produced a comprehensive Pavement Management Guide in 1977 [10]. Critical to both these efforts was recognition of the need to evaluate pavement behavior, pavement distress, and pavement performance steps in design. Prior to this time, design methods had attempted to predict performance directly from materials and weather inputs using empirical evidence such as the AASHO Road Test. These two initiatives in the late 1960s and early 1970s showed that it was essential to measure pavement behavior as an intermediate step since all known theoretical pavement equations do in fact predict behavior in the form of stress, strain, or deformation, but not performance directly. Behavior carried to its limit becomes distress in the form of cracking, permanent deformation, and disintegration. Distress as a function of accumulated traffic loads yields the required performance curve which can be used to judge the effective life of a pavement structure.

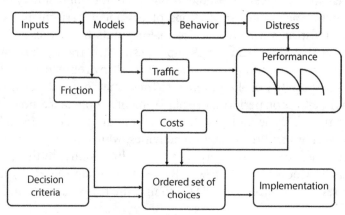

Figure 2.1 Major components of a Project Level Pavement Design System as initially formatted in the 1960s. (These remain true today.) After [9].

These factors properly analyzed can be used to determine required maintenance, overlay, and rehabilitation needs for the pavement including expected time of such interventions and the effect such interventions have on pavement performance life. The complete project level pavement management process then optimizes and compares predicted pavement performance life as a function of total life-cycle costs.

It quickly became clear that these same concepts of behavior and performance could be used to evaluate a group section in a pavement network by evaluating all the factors and developing performance prediction models for each individual section. In turn, the needs for each pavement in the entire set of pavement sections could be compared to determine when to intervene in each individual section and in what priority order to optimize budget expenditures and maximize total performance of the pavement network. All of these activities at both the project and the network level require data that defines the material properties, loads, environment, behavior, distress, and actual performance. The data must be stored in a central data base and be accessible to the entire pavement management process as illustrated in Figure 2.2. As well, the so-called feedback data that describes the actual performance of each individual pavement section of the many sections in the network must be accumulated and used to update the necessary performance and cost models as shown in Figure 2.3. Details about this basic process are described in [2,4,9]. We recommend them for study.

All of the earliest Pavement Management Systems described in the literature FPS (Flexible Pavement System), RPS (Rigid Pavement System), SAMP (Systems Analysis Method for Pavements), and OPAC (Ontario Pavement Analysis and Costs) operated at the project level and provided

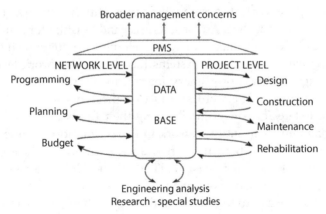

Figure 2.2 Components of a PMS, distinguishing the three levels. After [9].

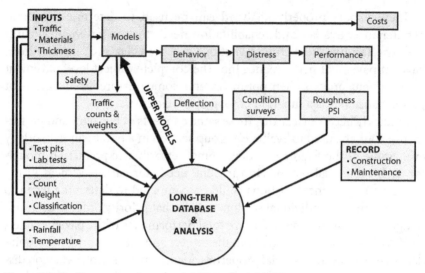

Figure 2.3 Performance and cost models diagram. After [9].

significant improvement in the ability to design, construct, and maintain pavements for adequate performance over time [4]. The functional performance and cost models used in all of these are represented in Figure 2.3.

2.1   Network Level PMS

At the same time many state and provincial DOTs were faced with trying to maintain and operate a large network of existing heavy duty pavements on the US interstate system and the Canadian National Highway System. Their concerns were with the many thousands of miles of the existing pavement network that were failing prematurely. This was the driving force for states such as Arizona, Kansas, and Washington, and several Canadian provinces to embark on the development of functional Network Level Pavement Management Systems [8,10]. Since network level pavement management was of primary interest to transportation executives and chief engineers, the word spread that these network level systems could assist states/provinces in allocating their funds in a more optimal way to maintain their entire network in better condition. These interactions occurred within both the American Association of State Highway and Transportation Officials (AASHTO) of the United States and the Transportation Association of Canada.

Other states began to follow suit and the growth of pavement management from the mid-1970s to mid-1980s advanced toward network-level programs. This was fostered by two National Workshops on Pavement Management sponsored by FHWA and AASHTO in which the Executive Directors of each organization attended.[1]

These workshops basically set the agenda for the next 10 years of development and buy-in of pavement management to other state/executives [11,12,13]. Moreover, this spreading interest led to the first conference in the series in Toronto in 1985 [14]. Because of the rapidly changing landscape of pavement management, a 1987 conference brought together representatives of most US state DOTs, Canadian provinces, and many local agencies, and led to the development of network level PMS in at least 50 of those 60+ agencies in attendance.

Those conferences were also attended by pavement engineers from other nations around the world, and they took the available PMS literature back home to develop their own network level PMS. At the same time undergraduate and graduate courses in pavement management were developed at the University of Texas by Dr. Ronald Hudson and at the University of Waterloo by Dr. Ralph Haas. These courses attracted both national and international students who formed the core cadre of working pavement management engineers around the world. In addition FHWA recognized a major need to reeducate practicing engineers who were working in transportation agencies before the advent of pavement management. In response to this need, FHWA funded an intensive pavement management six-week graduate level course at the University of Texas in Austin with Dr. Ronald Hudson as course coordinator. One hundred eighty-seven engineers from all over the world attended this course over the next three years. Drs. Ralph Haas and Matthew Witzack frequently served as guest lecturers in this course. The engineers attending from approximately 40 states and the FHWA engineers returned to their divisions and central offices and greatly expanded the quality and use of pavement management at the network level.

2.2   The Impact of Lack of Understanding of Software Requirements

As PMS evolved, a common mistake occurred: agencies attempted to develop a PMS in-house even though they lacked the computer and information

---

[1] The first workshop was held in Phoenix, Arizona, in May 1980, and the second in Charlotte, North Carolina, in September 1980. Drs. Hudson and Haas gave invited keynote addresses to the workshops.

technology skills, statisticians, and support staff to do the job properly. The problem resulted from a shortcoming in the education system of the universities that were teaching engineers in pavement management technology: educators <u>did not</u> adequately delve into the business processes and software needs associated with PMS. Some states and provinces, such as Arizona, Kansas, Minnesota, and Ontario, flourished by employing professionals in the disciplines to develop an appropriate network-level system with crude optimization. Unfortunately, the majority who tried to develop the systems in-house got bogged down or at best got a prioritization system based on worst first pavement condition across the pavement network. A number of these agencies in 2014 are still using these homegrown, self-developed pavement section prioritization methods that should not be called PMS at all. As we will discuss later, from 1997 to 2007, 20 to 30 state/provincial agencies advanced into better developed, more functional PMS with customized software from specialized venders. Similar activities included significant development in Chile as well as implementation of systems in two U.S. state DOTs, Brazil, Parana, Tocantins, and widely across Europe.

2.3   Lessons Learned from the Early Development Years

Pavement management had progressed in 20 years from a concept to a working process. The principles and definitions had been reasonably well formulated and much had been learned from implementation experience at Federal, State/Provincial, and local levels.

These twenty original or "early birth" years of pavement management experience (1967–1987) indicate that the original concept of a comprehensive, systematic process is quite valid. That is, it incorporates in an organized and systematic way all the activities that go into providing and operating pavements: they range from the collection, processing, and analysis of field and other data on various pavement sections; the identification of current and future needs; the development of rehabilitation and programs; to the implementation of these programs through design, construction, and maintenance.

What perhaps has not been so well learned is that just because an agency has carried out all these activities does not mean it has a functional PMS. To have a PMS in the proper definition of the term requires a coordinated execution of these activities, and most importantly, the incorporation of a number of key elements such as performance or deterioration models, lifecycle economic evaluation, etc.

Table 2.1 Some key ideas learned from 20 years (1967–87) of pavement management experience. After [15]

From P.M. Process Itself	From Using the P.M. Process
• The framework and component activities for P.M. can be described on a generic basis. • Existing technology and new developments can be effectively organized within this framework. • The framework allows complete flexibility for different models, methods, and procedures. • P.M. operates at two basic levels: network and project. • Sound technological base is critical to the process and its effective application to the process and its effective application.	• Development and implementation of a PMS must be staged. • Staging allows for understanding and acceptance by various users. • Options almost always exist; they should be evaluated on a life-cycle basis; this means we need models for predicting deterioration of existing pavements and rehabilitation or maintenance alternatives. • P.M. can make efficient use of available funds but it will not "save" a network if funding is below some threshold level. • Good information is essential to the effective application.

Table 2.1 summarizes, in more specific terms, some of the key ideas learned both about the P.M. process itself and about its application. These will be useful to reference during later enhancements in this book.

2.4   Basic Requirements for an Effective and Comprehensive PMS

Pavement management was determined in the early years to be no different in its general requirements than any other area of management. It involves the coordinated direction of resources and labor to achieve a desired end. Decision making at various levels is therefore a primary activity that can only be effective if good data/information is available.

Several additional basic requirements and businesses exist for the effective application of a PMS including the following:

1. Serving different types of users in the organization,
2. Making good decisions regarding network programs and individual projects, and executing these decisions in a timely manner,

Table 2.2 summary of activities and decisions within a complete pavement management structure. After [15].

Basic Blocks of Activities	Network Management Level (Administrative and Technical Decision)	Project Management Level (Technical Management Decisions)

A. Data	1. Sectioning
	2. Data
	  a. Field inventory (roughness, surface distress, friction, deflection, geometrics)
	  b. Other (traffic, unit costs)
	3. Data Processing
		1. Detailed data structural, materials, traffic, climate, and unit costs
		2. Subsectioning
		3. Data Processing

B. Criteria	1. Minimum serviceability, friction, structural adequacy, max. distress 2. Maximum user costs, maintenance costs 3. Maximum program costs 4. Selection criteria (max. of benefits and max. cost-effectiveness)	1. Maximum as-built roughness; minimum structural adequacy and friction 2. Maximum project costs 3. Maximum traffic interruption 4. Selection criteria (such as least total costs)
C. Analyses	1. Network needs (now) 2. Perf. Predictions and future needs 3. Maint. And rehab. Alternatives 4. Technical and economic eval. 5. Priority analysis 6. Eval. of alternative budget levels	1. Within-project alternatives 2. Testing and technical analyses (performance and distress predictions) 3. Life-cycle economic analyses
D. Selection	1. Final priority program of capital projects 2. Final maintenance program	1. Best within-project alternative (rehab. Or new pavement) 2. Maintenance treatments for various sections of networks
E. Implementation	1. Schedule, contracts 2. Program monitoring 3. Budget and financial planning updates	1. Construction activities, contract control, and as-built records 2. Maintenance activities, Maint. Management, and records

3. Making good use of the existing technology, and new technology as it becomes available,

4. More detailed discussion of these requirements is provided in [2,4,9].

An essential part of fulfilling the foregoing requirements is to have a structure or framework for the various activities of pavement management. Table 2.2 lists the major activities and/or decisions made within such a P.M. structure as summarized in 1987 [15].

Since 1987 we have made great strides in most of these areas and the result is greatly improved PMS software and a better understanding and implementation thereof.

3

Pavement Management Development from 2010

In 1987 the PMS process was generally being developed individually and in-house by state DOTs. All US states and Canadian provinces reported having some type of PMS in place. It has been estimated that of the 90% of these developed in-house, 60% were largely unused and 30% gave only simplistic answers based on condition. This growth of apparent PMS was spurred largely by the US FHWA mandate for pavement management (known as ISTEA) but lacked the guidance and available personnel and resources to truly succeed. Thus good systems were not always developed. Only in the order of 10% of the systems— including Arizona, Kansas, Washington, Minnesota, Alberta, Ontario, and a few others—were successful and used effectively.

At the time, it was not foreseen that private teams of engineers, system analysts, and programmers would recognize the need for effective user-friendly PMS and would step forward to work with several state/provincial DOTs to develop more complete PMS. Many did, and by 2010 approximately 35% of state/provincial DOTs were using commercial off-the-shelf systems successfully, three to four states per year were advancing their technology to improve PMS, and at least three or four agencies per year were adding other management systems (MMS, BMS, etc.), all working toward broader asset management.

In 1987 the subsequent need and development of asset management was also not adequately foreseen. But by 1997 it was receiving serious attention by AASHTO, TAC in Canada [16], FHWA, and in countries like Australia and New Zealand. In some states, asset management was used as an overarching planning tool. By 2010 several states recognized, however, that 90% of the assets to be managed were within the purview of PMS and BMS, pavements and bridges, and supported at the network level by maintenance management (MMS). Many had also expanded to fleet and safety management using commercial off-the-shelf systems [17].

3.1   Data Aggregation and Sectioning

The 1987 contribution also did not adequately foresee the need for improvements in data aggregation and PMS "sectioning." Since that time, individual PMS software vendors in contracts for specific state DOTs have devised sophisticated methods, including dynamic sectioning, for aggregating data that better represent sections or subsections of the pavement network under uniform conditions. This has been made possible by the fact that rapid network optimization analysis procedures have been developed which permit larger and larger networks of sections to be compared and optimized.

3.2   Private Investment

It is encouraging that several PMS providers in cooperation with their state DOT users have made significant investments in PMS software improvements in the last 10 years. While impossible to determine the exact amount, at least 20 state/provincial DOTs have invested approximately $3–4 million each in active pavement management. From this base the software providers have been able to spend significant funds on research and software improvements including clarifying the need for improved data collection methods. Although smaller in magnitude, these investments resemble the private sector investment made by Microsoft, Google, etc. to improve the software technology in their fields of endeavor. Of course, there remains a significant need for public investment in PMS research outlined in the FHWA Pavement Management Roadmap [18].

3.3   Parallel International Developments

There have also been significant parallel developments in the international community. Pavement management has flourished in Chile and to some degree in Brazil and other South American countries under the leadership of Dr. De Solminihac and other colleagues. Significant strides have been made to improve the properties and operating characteristics of the highway design model (HDM-4) under the leadership of the World Bank and carried forward by Drs. Kerali and Snaith, originally at the University of Birmingham, and others.

Significant developments have also been made in the United Kingdom and across Europe where a European PMS conference has been held several times in the last two decades. Pavement management has also spread to China and other Asian countries under the leadership of students who have learned their PMS in US/Canadian universities and returned to their home countries for application. Funding has been provided by such agencies as the Canadian International Development Agency (CIDA) and other nations. Time and resources available for preparation of this paper have unfortunately not permitted a more detailed listing and summary of these individual activities and references.

3.4   Administrative and Public Awareness of PMS

The advancement of asset management (AMS) remains an enigma. In fact, the Hudson, Haas, and Uddin 1997 book argued that Infrastructure Management was a better term than Asset Management or Facilities Management and thus chose it as the title [19]. General pursuit and sales of generic AMS concepts to state DOT administrators in many cases may actually have inhibited the use of PMS. AMS was sold as an overriding planning tool, vaguely outlining that all assets were to be combined and administered effectively. However, rigorous details of how this was to be done remain elusive. On the other hand, progress is being made from the bottom up. As of 2011, several states have adopted not only pavement management systems but have added maintenance management systems and bridge management systems. The combination of these three activities account, in most cases, for about 90% of the budget of state/provincial DOTs. See Figure 3.1. These systems also contain the data needed to do broader asset management. Several state DOTs are also adding safety management systems, fleet management systems, and facilities management

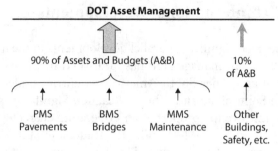

Figure 3.1 Components of Assets Management.

system. Combined with PMS, BMS and MMS, these management systems involve 98% of all the information and analysis needed for good asset management.

There is still need for coordination of the systems at the top, but this is occurring as administrators and agency business processes recognize the true value of bottom up information. Those agencies that are still trying to administer AMS from the top down are lagging in their use of effective management systems. This has occurred to some degree because AMS is sold by some to be a replacement or supplement for "planning." While planning is an important part of asset management, it can only function if the real data on facilities, pavement, bridge condition, and performance is available for analysis. In reality, planning is only one part of AMS.

3.5 Education

The continuing need for broader education in the pavement management field has not been fulfilled to the degree needed. Technical and analytical aspects of an effective PMS are broad and complex. Many DOTs do not have on their staff or even the ability to hire the disciplines needed, particularly statistics, economics, systems, and high quality computer programmers. Nor in general can they afford to develop or attract such employees to their normal staff. That may be best and most economically left to software providers/vendors who do have such personnel and who can apply the resulting technology over several agencies, thus reducing individual cost.

We also need to train existing DOT personnel more effectively. User-based education remains the great need across all state/provincial DOTs, cities, counties, etc. Stated another way, this is also an issue of knowledge management, which requires succession planning/continuity to be effective. Otherwise, like any asset, its value erodes.

The critical and ongoing need for education in PMS is well illustrated by the fact that while 200 practicing engineers trained in PMS from 1982 to 1984 at the University of Texas at Austin, over half took their new systems concepts and applied them in other fields and in turn were promoted to higher levels of responsibility in their agency within three or four years. As of 2011 more than 90% of those people had retired, leaving a major void in state DOT understanding of PMS. Thankfully, however, there are dozens of state DOT personnel who have seen the benefit of PMS and who have self-educated or taken appropriate short courses and/or worked with their PMS software providers to learn more fully the internal workings and benefits of PMS.

3.6   Improvements in Computers and Software Development

In the past 20 years there has been an order of magnitude improvement in PMS software and computers available to support it. In part, this was made possible by rapid advances in computer speed and low cost data storage capabilities. The software developments have been enhanced by a cadre of highly qualified analysts, statisticians, and software engineers who have been attracted into the field by the challenge and the funds made available by software entrepreneurs who have invested in software that they now vend to various state and provincial agencies.

3.7   Other Compatible Management Systems

In 1987 there was a general indication of the broader interfaces under an asset management system (AMS) umbrella. However, what was not foreseen was the increased development and use of modern maintenance management (MMS) which in many states/provinces led the way to later implementation of pavement management. At least 8–10 states/provinces that now use strong PMS started after an active MMS whetted their appetite for high-speed data processing, optimization, and decision making. The success and interface with MMS led those agencies to move more rapidly into PMS and to integrate the two.

3.8   Expansion of PMS Concerns

All of us are beginning to see an expansion of PMS concerns that include noise, societal, and environmental aspects. Terms like "sustainable pavement management" and "sustainable pavements" have been gaining traction, but rigorous definitions are still lacking. The ideas have merit and generally seem to mean trying to produce pavements with greater concern for societal effects such as noise, user costs, user delays, etc. and environmental factors such as consideration of hydrocarbon output, carbon footprint, global warming, etc. Progress is needed in these areas.

No one ever proposed that management systems be used to replace a good estimate of initial design. Indeed the concept has always been to develop the best possible initial design with available inputs and within reasonable budgets, but we must also accept the fact that no matter how well we design, Mother Nature and statistical factors will change in the 20, 30, or 50 years after the initial design and these must be taken into account with management systems.

4

Setting the Stage

Part Three describes the logical next step using data to determine needs and priority programming, rehabilitation, and maintenance. This requires establishing criteria to identify needs deterioration modeling for alternative rehabilitation and maintenance treatments, cost, benefit analysis, and priority programming methodologies. Examples are provided to illustrate the activity.

Part Four describes the Framework and Methodologies for project level design. This involves structural and life cycle economic analysis of available flexible and rigid pavement alternatives. It gives more detailed physical, cost, and other design inputs, the actual analysis models used, and example applications with particular attention to the current Mechanistic Empirical Pavement Design Package. While pavement design is a key part of pavement management, it is stressed in Part Four that good design is not enough by itself; good pavement management has to be practiced as a total process.

Part Five presents a logical sequence of implementation phases in overall pavement management. The steps involved are first defined and then the prominence of software providers is identified. Pavement preservation is described as a key component of pavement maintenance. Since pavement management is implemented within the broader context of asset management, the issues involved are also addressed.

Examples of Working Systems, at both the network and project levels, as described in Part Six, illustrate how pavement management systems are used in practice. There has been an evolution in development and application but basic features of working systems remain constant. Major change in the evolution has been the replacement of in-house development with use of vendors who provide comprehensive software packages. Examples of prominent vendors are given in Part Six. As well, HDM-4 is largely done by consultants. Comprehensive development of city or municipal PMS over the past two decades is noted as the implementation of airport PMS.

Looking ahead is an essential feature of good pavement management. The authors feel that this is still an entirely essential feature and is the focus of Part Seven. The section covers the use of PMS to solve special problems as well as the need to integrate new technologies as they emerge. Although PMS has evolved to a full-function, it is not complete or perfect. Part Seven identifies still needed elements. Finally, the way that PMS has led the way to functional asset management is briefly covered in the final chapter.

The more the engineering community can understand and truly accept Management Systems as the required methodology for the variable real world, the faster we will make progress.

References for Part One

1. Roads and Transportation Association of Canada, "Pavement Management Guide," Ralph Haas Editor, RTAC, Ottawa, 1977.
2. Haas, Ralph and W.R. Hudson, *Pavement Management Systems*, Kreiger Press, Malabar, Florida, 1978.
3. Transportation Association of Canada, "Pavement Asset Design and Management Guide," Susan Tighe Editor, TAC, Ottawa, Canada, 2103.
4. Haas, Ralph, W.R. Hudson, and J.P. Zaniewski, *Modern Pavement Management*, Kreiger Press, Malabar, Florida, 1994.
5. Yoder, E.J. and M.W. Witczak, *Principles of Pavement Design*, 2nd Ed., Wiley, 1975.
6. NCHRP, "Automated Pavement Distress Collection Techniques," *NCHRP Synthesis 334*, Transportation Research Board, Washington, D.C., 2004.
7. Uddin, Waheed, W.R. Hudson and R. Haas, *Public Infrastructure Asset Management*, Second Edition McGraw Hill, 2013.
8. Hudson, W.R., F.N. Finn, B.F. McCullough, K. Nair, and B.A. Vallerga, "Systems Approach to Pavement Systems Formulation, Performance Definition and Materials Characterization," Final Report, NCHRP Project 1-10, Materials Research and Development, Inc., March 1968.
9. Hudson, W.R., and B.F. McCullough, "Flexible Pavement Design and Management Systems Formulation," NCHRP Report 139, 1973.
10. Haas, Ralph (Editor), RTAC, "Pavement Management Guide," Roads and Transportation of Canada, 359 pp, Ottawa, Canada, 1977.
11. Hudson, W.R., S.W. Hudson, and P. Visser, "Measuring Benefits Obtained from Pavement Management," published *Proceedings*, 5th Intl. Conference on Managing Pavements, Seattle, August 2001.
12. Hudson, W.R. and R. Haas, "Maximizing Customer Benefits as the Ultimate Goal of Pavement Management," Published *Proceedings*, 5th Intl. Conference on Managing Pavements, Seattle, August 2001.
13. Hudson, W.R., L.O. Moser, and W. Visser, "Benefits of Arizona DOT Pavement Management System after 16 years Experience," published *Proceedings*, 7th Conference on Asphalt Pavements for Southern Africa, CAPSA 99, Victoria Falls, Zimbabwe, August 29, 1999.
14. First North American Pavement Management Conference, *Proceedings*, Ontario Ministry of Transportation and Communications, March 1985.
15. Haas, Ralph and W.R. Hudson, "Future Prospects for Pavement Management," *Proceedings*, Second North American Conference on Managing Pavements, Ministry of Transportation Ontario, Toronto, November 1987.
16. Transportation Association of Canada, "Pavement Design and Management Guide," TAC, Ottawa, Canada, 1997.
17. Hudson, Stuart and W. R. Hudson, "Improving PMS by Simultaneous Integration with MMS," paper presented at the 8th ICMPA 2011, Santiago, Chile, 2011.

18. Zimmerman, KA, L. Pierce, and J. Krstulovich, "A 10-Year Roadmap for Pavement Management," FHWA Executive Report #HIF-11-014, December 2010.
19. Hudson, W.R., R. Haas, and W. Uddin, *Infrastructure Management*, McGraw-Hill Publishers, New York, 1997.

Part Two
DATA REQUIREMENTS

5

Overview of Pavement Management Data Needs

5.1 Classes of Data Required

An engineering-based focus on the condition, structural capacity, performance, and safety of existing pavements is necessary. But it must be supplemented with information concerning the inventory of the network and data related to environment and policies, which affect management decisions.

The various classes and component types of pavement data are applicable to either rehabilitation or maintenance decisions. Modern terminology would describe the maintenance areas as components of *pavement preservation*. This shift in terminology suggests that the management of existing pavement assets is a growing concern as the highway infrastructure throughout the world has aged.

The data base is a central feature of a PMS in the modern context. With continuing advancements in computer and software technology, there have been improvements in the amount of data that can be stored and readily accessed, but the role of the information in the data bases has not been diminished.

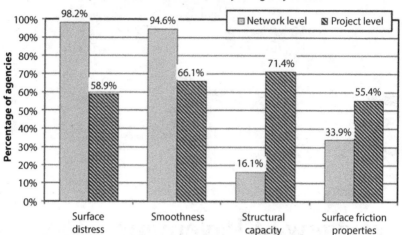

Figure 5.1 Types of pavement condition data collected for pavement management systems. After [1]

The types of data needed for pavement management can be broadly classified as inventory data and pavement condition data. Data describe the relatively permanent features related to the pavement sections and would typically include pavement construction and maintenance records, geometric features, etc. Traffic data were typically associated with the inventory files as information about traffic volumes and truck loadings were only periodically updated. The pavement condition data include measures of pavement quality, which are broadly classified as performance/roughness, structural, distress, and safety/skid data. Figure 5.1 shows the distribution of the types of data collected by state highway agencies for project and network level applications [1].

The type of data needed depends on the uses of the information at both the network and project-levels, as summarized in Table 5.1 [2]. A more comprehensive and still relevant listing of performance related, historic related, policy related, geometry related, environment related, and cost related uses of data is provided [3].

5.2   The Importance of Construction and Maintenance History Data

The importance of construction and maintenance history data has been recognized for several decades. Significant issues in the early 1990s were isolation of pavement management systems from other functional areas

Table 5.1 Network- and project-level data collection. After [2]

Aspect	Network-level	Project-level
Uses	• Planning • Programming • Budgeting • Pavement management system treatment triggers, identification of candidate projects, life cycle cost analysis • Network-level condition reporting • Mechanistic-Empirical Pavement Design Guide (MEPDG) calibration	• Project scope • Refine pavement management system treatment recommendations • MEPDG calibration
Data Items Typically Collected	• IRI • Rut depth • Faulting • Cracking • Punchouts • Patching • Joint condition • Raveling • Bleeding • Surface texture	• Detailed crack mapping and other distresses • Structural capacity (e.g., falling weight deflectometer [FWD]) • Joint load transfer • Base/soils characterization (e.g., ground penetrating radar, cores, trenches)
Other Items Collected Concurrently	• Video • GPS coordinates • Geometrics (e.g., curve, grade, elevation, cross slope) • Other assets (e.g., bridges, signals) • Events (e.g., construction zones, railroad crossings)	• Drainage conditions • Appurtenances (e.g., sign and guardrail location and condition) • Geometrics (e.g., curve, grade, elevation, cross slope)
Speed	• Typically highway speeds	• Walking or slower speeds

of highway agencies and the lack of integration of a central data base that captured construction and maintenance information. With the evolution to enterprise management systems for many highway agencies, as well as other functions of government, there is greater potential for capturing agency wide information needed within the focus of the pavement management system.

5.3   The Importance of Performance Related Pavement Evaluation

Performance related pavement evaluation is directed primarily to how well the pavement is providing serviceability to the user, what level of surface distress exists, the existing structural adequacy, and the level of safety in terms of surface friction. These four measures, along with maintenance and user costs, can be viewed as "outputs" of the pavement.

Performance related evaluation is increasingly becoming important in performance management, particularly in terms of selecting performance indicators [4].

5.4   Objectivity and Consistency in Pavement Data Acquisition and Use

Consistent and repeatable quantification of pavement data is an essential requirement. This has to occur over time and space: proper training is needed and a set of well documented practices and procedures is also necessary.

5.5   Combining Pavement Evaluation Measures

Combined or aggregated measures for overall quality or condition of a pavement section, or a network, are particularly useful to senior management levels. This is further discussed in Chapter 11.

6

Inventory Data Needs

6.1 Purpose of Inventory Data

The inventory of existing facilities is an essential feature of PMS. It includes section location, geometry, pavement structure, costs, environmental data, and traffic. That requirement has not changed. However, the methodology for populating and managing the inventory has changed significantly with technological advances.

6.2 Types of Inventory Data

The general types of inventory data listed in Table 5.1 and noted above are relevant to a PMS. However, with the expansion of management systems to the entire infrastructure of an agency and even to the enterprise level of governmental agencies, the ability to integrate data bases is more vital than ever to modern management systems.

6.3   Selection and Referencing of Pavement Management Sections

There are alternative techniques for establishing a reference method and sections for PMS. In the early 1990s, Geographic Information Systems (GIS) were in their infancy. However, at that time the use of GIS was identified as an important future technology for pavement management. This intuition was correct and it is difficult to envision any modern PMS that does not use GIS.

GIS is an important technology in almost every area of activity. There is a wealth of information on the subject, including textbooks [5,6] and many other technical articles on transportation. There are numerous descriptions of applications, such as that in Utah [7]. Any attempt to provide a thorough description is beyond the scope of this book. However, with respect to a pavement management application, the following key features are relevant:

GIS combines geographical features with tabular data. The geographical features are essentially digital maps with the special data points and features defined by global coordinates [8]. For PMS the GIS maps are generally established using linear referencing, such as defined below by ESRI [8], a major vendor of GIS software:

> A method for storing geographic data by using a relative position along an already existing line feature; the ability to uniquely identify positions along lines without explicit x, y coordinates. In linear referencing, location is given in terms of a known line feature and a position, or measure, along the feature. Linear referencing is an intuitive way to associate multiple sets of attributes to portions of linear features.

Historically, establishing the GIS "map" was labor intensive. However, GIS maps are now readily available from many sources. The association of multiple data sets to portions of the linear features is a key GIS technology. PMS data sets with different attributes, such as data bases for construction history, material properties, condition, accidents, etc. are associated with the linear references, even though the original structure of the data base may have been referenced based on construction project location, milepost referencing for condition data, or mile-point referencing for accident data. Linear referencing provides a common methodology for identifying the roadway location of each element in various data bases.

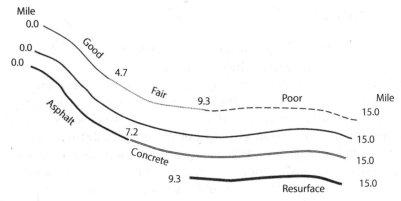

Figure 6.1 Simplified example of dynamic segmentation concept. After [9]

Implementation of GIS brought the ability to look across data bases to identify the management sections and their attributes. GIS uses the defined PMS sections and *dynamic segmentation* to identify the attributes specific to each section. Dynamic segmentation is defined in [8] as:

> The process of computing the map locations of linearly referenced data (for example, attributes stored in a table) at run time so they can be displayed on a map, queried, and analyzed using a GIS. The dynamic segmentation process enables multiple sets of attributes to be associated with any portion of a line feature without segmenting the underlying feature. In the transportation field, examples of such linearly referenced data might include accident sites, road quality, and traffic volume.

Figure 6.1 shows a simplified example of the concept of dynamic segmentation. Four attribute data bases are mapped onto a common scale: condition (top line), traffic (second line), pavement type (third line) and rehabilitation (bottom line) [9].

6.4   Collecting and Processing Section and Network Data

Modern technology based on Global Positioning Systems (GPS) can be used to collect other inventory information. Frequently, these capabilities are included in an integrated survey vehicle that is also capable of collecting pavement distress and roughness data. Integrated survey vehicle examples are subsequently described in Chapter 9.

6.5   Traffic and Truck Load Data

Empirical pavement design methods were predominantly used for highway pavements for decades. Relevant traffic data could be summarized with Annual Average Daily Traffic (AADT) and the number of equivalent single axle loads on a section. With the pursuit of mechanistic-empirical design methods, better truck loading information is needed, such as the spectra of truck axle loadings. Traditionally, static scales were used for truck weight measurements. While these scales are accurate, the data quality suffers from limited operational times, which can lead to issues with not measuring a representative sample of the truck population.

Weigh in Motion (WIM) technology has been implemented by various agencies to capture truck weight and axle load data. A WIM station basically uses a sensor embedded in the pavement and associated hardware and software to measure truck and axle weights at either low or high speeds. Low speed WIMs are used as a screening enforcement tool. High speed WIMs are used for monitoring truck data for engineering applications such as pavement design and management and/or screening trucks approaching a weigh station.

Sensor technologies include load cells, bending plates, piezoelectric, and quartz. The bending plate and piezoelectric have been widely implemented. One of the most comprehensive records of traffic load data collection and sensor technologies resides in the LTPP (Long Term Pavement Performance Program), and extensive descriptions of data, equipment, methodologies and procedures, etc. are provided in LTPP InfoPave (www .fhwa.dot.gov/research/tfhrc/programs/infrastructure/pavements/ltpp/).

7

Characterizing Pavement Performance

7.1 The Serviceability-Performance Concept

The pavement serviceability-performance concept provides valuable information for pavement engineers, managers, and administrators. As developed at the AASHO Road Test [10], the concept has its foundation in the users' perception of the quality of service provided by the pavement at one point in time.

7.2 Pavement Roughness

The primary variable influencing serviceability is the roughness of the surface. Roughness is characterized by distortions in the pavement surface that contribute to an undesirable or uncomfortable ride. More specifically, the component of the pavement's surface most directly related to roughness is the distortions in the longitudinal wheel paths. These distortions generate a vertical acceleration to the passenger compartment that is perceived as uncomfortable by the user, depending on the frequency and amplitude. Roughness evaluation is extremely important to the network PMS process as it provides the pavement manager with a direct measurement

influencing the public's perception of the quality of service provided by the pavement. Moreover, as the technology for measuring profile has evolved, additional applications of pavement profile data have come into use, notably for construction quality control and as input to pavement design/construction. For construction quality control, precise measurement of roughness is needed to help define contractor pay factors. In the pavement design process, profile data can be used to estimate the minimum overlay thickness needed to achieve a desired level of post-construction smoothness.

Alternatively, template correction strategies can be used, such as milling to increase the smoothness of the existing pavement surface prior to overlay.

Three components are needed for pavement roughness evaluation:

- An accurate measurement of profile.
- A mathematical model, or filter, to convert the profile data into a meaningful summary statistic that captures the important distortions in the profile that relate to the use of the statistic.
- An interpretation of the roughness statistic.

7.3   Equipment for Evaluating Roughness

Technology in roughness evaluation equipment has focused on the measurement of the pavement profile. The response type road roughness measuring systems are largely related to legacy applications so only advances in profile measuring are discussed herein.

Some agencies continue to use profilographs for construction quality assurance, particularly for concrete pavements. The discussion of these devices in [3] remains valid so they are not discussed further herein.

Four classes of profile measuring equipment have been defined to meet the current needs of profile measurements:

- High speed profilometers – used for network monitoring.
- Light weight profilometers – developed for construction monitoring.
- Walking profilometers – used for validation and application for construction monitoring.
- Conventional precision surveying methods.

Under the SHRP 2 research program, measuring smoothness during the construction of concrete pavements, while the concrete is still wet, is being evaluated. The initial report on this topic found the concept is viable but the available technologies are not adequate for implementation [11].

High-speed measurements are necessary for monitoring pavement networks. The current state of the practice is to measure profile with devices using the principles of inertial profilometry, including the French APL as still used in Europe [12]. With the increased understanding of the need and application of profile data, there are several profiling devices available on the market in 2014. The evolution of inertial profilometers has produced equipment with different sensors, filters, and calculation approaches from multiple manufacturers. Due to the dynamics of the industry, a specific review of the capabilities of the currently available devices would not serve the needs of the reader. The market should be assessed at the time of making an equipment selection decision. However, for the foreseeable future the key issue is the reliability of the measurements. In that sense the reader is referred to the profile measurements in LTPP, as detailed in www.infopave. com with regard to equipment, results, quality assurance, and calibration.

7.4   Toward a Universal Roughness Standard

A universal roughness standard has been the subject of extensive discussion. In the last two decades the International Roughness Index (IRI) is widely used internationally. It is the statistic of choice of the U.S. Federal Highway Administration and is therefore used by many state highway agencies. However, it must be recognized that IRI is a measurement unit for pavement roughness, much as the meter is a measurement unit for length. Knowing the IRI of a pavement has little meaning without an understanding of what the magnitude of the number means. Figure 7.1 attempts to provide a contextual meaning of the magnitude of IRI [13]. In the metric system, IRI is expressed in m/km. In the U.S. the units of inches/mile are used. The factor to convert from metric to U.S. is 63.36, e.g. on Figure 7.1 value of 2 mm/km would equal 127 inches/mile.

Whatever scale is established, the user of the measure must still make decisions as to what is an acceptable level of roughness. Most agencies differ in this important aspect and that means the IRI is far from universal in actual use. Table 7.1 compares the Wisconsin DOT (WisDOT) and FHWA assessments of pavement quality based on IRI [14]. The reasons for these differences are explained in the WisDOT report as:

> There is a difference between WisDOT IRI categories and the Federal
> Highway Administration (FHWA) categorization of IRI. This is

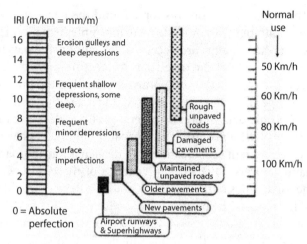

Figure 7.1 Contextual illustration of magnitude of IRI measurements. After [13]

Table 7.1 Different interpretations of pavement quality based on IRI. After [14]

IRI Categories of Roughness	WisDOT	FHWA
Very Good	≤ 95	≤ 60
Good	96–170	61–95
Fair	171–220	96–120
Poor	221–320	121–170
Very Poor	> 320	> 170
	Unacceptable cells	

because each uses IRI for a separate purpose. WisDOT uses its IRI measurement as a roughness index for the purposes of programming projects. WisDOT also uses a pavement cracking index and a rutting index to indicate when a road should be resurfaced. The vast majority of time a section of roadway will need rehabilitation based on cracking or rutting. Roughness is usually a "lagging" indicator that shows the road is rough after other problems (like cracking and rutting) have become severe. On the other hand, the FHWA categories of IRI were originally developed for Interstate Highways. FHWA uses IRI as a performance evaluation tool, especially for comparing relative performance state to state. Thus the WisDOT and FHWA IRI Categorizations are scaled to fit a different purpose.

The difference in roughness categories shown between WisDOT and FHWA illustrates the problem with IRI. A "very good" rating in WisDOT has little meaning compared to the FHWA scale and similar variation exists among most states. The Present Serviceability Index (PSI) developed at the AASHO Road Test [15] is still a useful, stable method of comparison across all agencies because it is an ordinal scale of 0–5 and evaluates human response to roughness in a uniform way.

7.5   Calibration Needs and Procedures

Proper calibration of inertial profiles is essential. The following discussion is directed primarily to that subject.

The high speed and light weight profilometers use principles of inertial profilers. While the theory of inertial profilometry is fundamentally sound, there is a plethora of issues related to the instrumentation, profile sampling, analysis methods, and operation of profilometers that affect both the precision and bias of the measurements. The technical details of specific equipment are of less concern than the ability of the equipment to provide reliable roughness measurements. Both AASHTO and ASTM have specifications and test methods for measuring profiles. The relevant AASHTO specifications are:

- M 328-10 Standard Specification for Inertial Profiler
- R 54-10 Standard Practice for Accepting Pavement Ride Quality when Measured Using Inertial Profiling Systems
- R 56-10 Standard Practice for Certification of Inertial Profiling Systems
- R 57-10 Standard Practice for Operating Inertial Profiling System

The proliferation of profilometer makers and users created concern with the accuracy and precision of profile data. The FHWA initiated a pooled fund study, "Improving the Quality of Pavement Profiler Measurements" TPF-5(063), in 2002. See also the flyer the FHWA used to promote the study [16].

The essence of the process supported by a pooled fund study is described in [17]. Again the LTPP program (www.infopave.com) also has comprehensive up-to-date information.

As shown in Figure 7.2, profilers can be evaluated based on a roughness summary statistic or based on comparison of the profiles measured by the

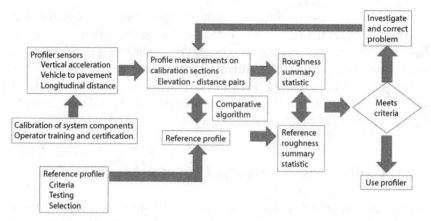

Figure 7.2 Profiler calibration and verification process. After [17]

profiler to reference profiles. Three things are needed for the calibration/ verification process:

- Measurement of "true" reference profile.
- An analytical process for comparing the reference profile to the data from the profiler.
- Criteria for determining acceptable bias and accuracy between the profiler and the reference results.

One of the tasks in the pooled fund project was to compare the different profiling equipment. Table 7.2 identifies the equipment that participated in the study [17,18]. It also demonstrates the range of available equipment types in 2002. Details on manufacturer, source, etc. are contained in [17,18]. As well, readers can find many of these details by simply entering the make and/or model in an internet search.

Profile data were collected on five sites: two smooth asphalt surfaces, a rough asphalt surface, a continuously reinforced concrete pavement, and a jointed concrete pavement. The repeatability and bias of 19 of the 68 devices that participated in the roundup are provided in [18]. Results for the jointed concrete pavement were not included over concerns that movements at the joints between measurements were potentially causing changes in the profile that would affect the comparisons of the equipment. This is not a valid concern but is a widely held misunderstanding. Rod and level surveys were used to establish the reference profile. The IRI was used as the summary statistic.

During the pooled fund study, the high-speed profilers were evaluated for survey and construction monitoring data collection; the light-weight

Table 7.2 Profiling devices participating in FHWA pooled fund study. After [17]

Class	Profiler Make/Model	Number in Study
High-Speed	ROSAN	3
	ICC	13
	ARAN	6
	MGPS	1
	Custom	3
	RSP five0five1	2
	ROADMAS	1
	Pathway	2
	K.J. Law	2
	MHM	2
	Digilog VX	1
	Starodub/DHM	1
	SSI	1
	Ames	1
Light-Weight	Starodub/ULIP	1
	ICC	6
	SSI	4
	Dynatest/Law T64	1
	K.J. Law	2
	Custom	1
	Transtology	1
	Ames	2
Walking-Speed	SuPro 1000	3
	R/D-Meter	2
	ARRB WP	3
	Rolling Rod and Level	1
	YSI RoadPro	1
	COMACO GSI	1
	ROADMAS Z2	1

profilers for construction monitoring; and the walking profilers for reference. Repeatability and bias analysis determined that several of the high-speed profilers showed acceptable performance on the smooth sections, indicating they could be used for construction monitoring. However, several others did not meet the criteria for the rough asphalt section, which brings into question their value as survey devices.

While the IRI statistic is a well-defined roughness summary statistic, profiler manufacturers take some latitude in the processing of the profile data, which can result in the software from the different manufactures producing different IRI results from the same profile data. To overcome this problem, Transtec, under contract with the FHWA, developed the ProVal software [19]. This software is available for free and provides a common methodology for computing IRI. Per the ProVal web-page, the following can be computed from the profile data:

- Standard Ride Statistics
- Fixed-Interval Ride Statistics
- Continuous Ride Statistics
- Power Spectral Density (PSD)
- Profilograph Simulation
- Rolling Straightedge Simulation
- Localized Roughness Identification (Tex-1001-S); (Version 2.7 and earlier)
- Cross Correlation
- Profiler Certification
- Precision and Bias (ASTM E 950)
- Smoothness Assurance Module (SAM)
- Automated joint Fault Measurement (AFM)
- Optimal Weigh-in-Motion Site Locator (OWL)

As shown on Figure 7.2, the actual profiles from profilers should be compared to reference profiles. This requires an analytical method for comparing two profiles and criteria for determining if the profiler measurements are acceptable. There are two methods used for this comparison, power spectral density (PSD) and cross-correlation. The PSD is used to identify the "wavelength" component of the pavement profile. The profiles can then be processed with filters to separate out the wave lengths, or wavebands of interest, which for critical profile accuracy requirements are [20]:

- Short waveband: The profile, passed through a high-pass filter with a cutoff wavelength of 1.6 m (5.25 ft) and a low-pass filter that is customized to reproduce the bridging and filtering applied by the reference or benchmark profiling device.

- Medium waveband: The profile, passed through a high-pass filter with a cutoff wavelength of 8 m (26.2 ft) and a low-pass filter with a cutoff wavelength of 1.6 m (5.25 ft).
- Long waveband: The profile, passed through a high-pass filter with a cutoff wavelength of 40 m (131.2 ft) and a low-pass filter with a cutoff wavelength of 8 m (26.2 ft).

The cross-correlation analysis can then be used to identify the extent to which the profile from a profiler mimics the reference profile. The analysis methods were evaluated and the mathematics for PSD and cross-correlation were developed in the pooled fund study [20]. Recommendations from this report are embedded in the ASTM and AASHTO profiler calibrations and verification standards. The calculations are used in the ProVal software.

As part of the pooled fund study, potential "benchmark" profilers were evaluated in 2009 and 2010 [21]. There was a marked improvement in the equipment between the two surveys, demonstrating the dynamic nature of developments in roughness evaluation. Only the 2010 results are presented in the following.

The testing included five pavement sections at the Minnesota Road Research Facility in Albertville, Minnesota, and one section of U.S. 10 near Junction City, WI. Macrotexture type and smoothness were dominant criteria for selecting test sections. The surface types included:

- Dense graded asphalt (DGA),
- Fresh chip seal (CS),
- Pervious hot mix asphalt (PHMA),
- Transversely tined concrete (TT), Longitudinally tined concrete (LT),
- Diamond ground concrete (DGC).

The candidate reference profilers in the 2010 experiment were:

- Dipstick 2000 [22]
- APR Auto R&L [23]
- SSI CS8800 [24]
- SurPro 3000+ [25]
- VTPL TMS [26,27]

The first four devices are commercially available. The SurPro device was the model 3000+, which International Cybernetics is now marketing as model 3500 with enhanced capability. The VTPL TMS is a custom terrain measuring system developed at Virginia Polytechnic Institute and State University (Virginia Tech).

Under the pooled fund study, a Benchmark Profiler [28] was developed at the University of Michigan to provide a highly accurate, highly repeatable instrument. Profiles from the Benchmark Profiler were used for the evaluation of the candidate reference profilers with respect to accuracy, repeatability, and longitudinal distance measurement, as summarized in [21]. The researchers concluded:

> None of the candidate reference profiles achieved a passing score on all of the criteria. The SurPro 3000+ achieved passing repeatability and accuracy scores for profile over the broadest range of conditions in the 2010 experiment, and it achieved a passing repeatability score for all but the short waveband in nearly every case. However, the unit did not achieve passing scores for longitudinal distance measurement on all of the test sections.

Even though the pooled fund study is ongoing, some of its results are being implemented. The Minnesota DOT has adopted the SurPro as the reference device for their calibration and verification of high-speed profilers [29]. An Oregon study also found the SurPro measures highly repeatable and accurate profiles [30].

As indicated in their Pavement Distress Identifications Manual [31], their regular high speed profilers are part of the Pathway Services Inc. Digital Inspection Vehicle with lasers for longitudinal profile and ruts and four digital cameras for distress.

During the pooled fund study, careful attention was paid to the calibration and verification protocols. It can be expected that in a less carefully controlled environment there would be greater discrepancies in the measurements between agencies and across time. An NCHRP study of comparative performance measures [32] across states noted that "the analysis conducted within this project was performed with the understanding that current variations in IRI measurement practice make precise comparisons of IRI across states (or even across survey efforts within a state) difficult" [32].

While major progress has been made in the evaluation of pavement roughness, it is expected that as the pooled funded project reaches fruition even further progress will be made. The following recommendations were presented in the report (the recommendations are quoted, except current AASHTO standards were substituted for the provisional standards cited in the report [32]):

1. Encourage adherence to the AASHTO M 328-10 Standard Equipment Specification for Inertial Profilers. Use of a recording interval less than two inches is a key element of this specification that is not yet standard practice because it can require expensive modification of existing profiler components.
2. Encourage rigorous application of regular calibration procedures and system checks, as documented in the AASHTO R 57-10 Standard Practice for Operating Inertial Profilers and Evaluating Pavement Profiles. Most importantly, use regular equipment calibration and daily system checks to ensure integrity of network IRI surveys.
3. Further develop AASHTO M 328-10 and R 57-10 for network profilers. These standards were written with construction quality assurance in mind and can be improved based on current experience with network profiler application. Consideration should be given to adding specifications for real-time data quality checks.
4. Spot check profile data on control sections to ensure that profilers are functioning properly.
5. Verify IRI calculation software—wherever software is used to generate IRI values, they should be verified using a reference program. This is best accomplished via a collective effort involving profiler manufacturers.
6. Require profiler accuracy and repeatability testing as a condition of procurement contracts. Certify existing profilers against a defensible reference measurement, and upgrade them as needed.

7.6   Relating Roughness to Serviceability

The serviceability concept as developed at the AASHO Road Test [10] was to have a pavement performance measure that was related to the users' opinion of the quality of service that a pavement provided at one point in time. A scale of 0 to 5 was selected so users could relate their experience in a simple format. It is not possible for users to rate pavements on the IRI scale. It is also difficult for non-technical personnel, such as the public or administrators, to understand the magnitude of values on the IRI scale.

Due to the importance of the serviceability concept, when the IRI was introduced there was a lot of interest in correlating IRI with serviceability. One of the earliest efforts was by Al Omari and Darter [33] who correlate PSR, the mean of user ratings on a 0 to 5 scale, with IRI to establish the following regression relationships for metric and U.S. customary measurement units:

$$PSR = 5 * e^{(-0.26*IRI)}$$ where IRI is in millimeters per meter.
$$PSR = 5 * e^{(-0.0041*IRI)}$$ where IRI is in inches per mile.

Another conversion is provided in the 1997 Canadian Pavement Design and Management Guide [34], as follows:

$$RCI \text{ (Riding Comfort Index)} = 10*C^{-0.18\,IRI}$$

Where RCI is on a scale of 0 to 10 and IRI is in mm per meter.

Another issue with using IRI as a surrogate parameter for users' opinions is that the IRI statistic was developed for the calibration of Response Type Road Roughness Measuring Systems (RTRRMS) equipment. The quarter car parameters that filter the profile in the quarter car model were selected to represent the measurement capability of these devices.

Unfortunately, this is not the best filter for a profile statistic that has high correlations with user ratings. Sayers and Karamihas [35] introduced the Ride Number (RN) concept to address this issue. RN is computed as an exponential transformation of a roughness summary statistic, termed Profile Index (PI) as:

$$RN = 5e^{-160(PI)}$$

The Profile Index uses a quarter car filter of the profile data similar to IRI, but with the following differences:

- The IRI coefficients are replaced with $K_1 = 5120$, $K_2 = 390$, $C = 17$, and $\mu = 0.036$.
- The initialization length was increased from 11.0 m (IRI) to 19.0 m (PI).
- The accumulation is done by root-mean square, rather than mean absolute.

7.7   Applications of Roughness Data

The use of roughness and other data for network level and for project based purposes is indicated in Table 5.1 of Chapter 5. Many urban areas now commonly do roughness surveys, as subsequently described in Part Six, Examples of Working Systems. Another widespread use of roughness evaluation at airports is described in Part Six.

8

Evaluation of Pavement Structural Capacity

8.1   Basic Considerations

Pavement structural evaluation can be broadly classified as nondestructive or destructive. The data acquired is essential to assessing the structural capacity of pavement sections and networks. In the former case, it is commonly used in the design of rehabilitation treatments such as overlays, and for networks it can involve the allocation of funds.

8.2   Nondestructive Measurement and Analysis

The introduction and data collection plan described in [3] remains valid and should be reviewed. Deflection measurements remain the primary method of nondestructive structural evaluation. Effective use of deflection data requires knowledge of pavement layer thicknesses. Ground Penetrating Radar (GPR) [36–44] is a nondestructive option for determining pavement thickness, as are as-built construction records. GPR is further discussed in Section 8.2.3.

8.2.1   Deflection Measurements

The Falling Weight Deflectometer (FWD) is the most widely used device for measuring pavement deflection. Alavi *et al.* synthesized the state of highway practice for using deflections for NCHRP [45]. Table 8.1 summarizes the responses of 45 state highway agencies concerning the brand and age of the equipment of the FWDs they use. It can be noted that the Dynatest [46] is the most common FWD, with 61 in use, but the average age is 14 years. The JILS [47] equipment is the second most common with 15 in use, and the average age is much shorter at six years. The KUAB [48] devices are the least common with only six in use and an average age of 14 years.

With multiple number and brands of devices, the need for compatible measurements within and across agencies becomes evident. The Long-Term Pavement Performance (LTPP) program has emphasized the importance of calibration of FWD devices. Standard calibration methods were developed with the most recent iteration prepared by [49] as part of a FHWA Transportation Pooled Fund Study 5(039), initiated in 2004. This study initially set up calibration centers in Pennsylvania, Texas, Minnesota, and Nevada (subsequently moved to Colorado). These have been augmented with centers in California and Montana. All of these centers are operated by state highway agencies. In addition, manufacturers have set up calibration centers: Dynatest in Florida, Foundation Mechanics (JILS) in California, and Carl Bro in Denmark. Additional international calibration centers have been established in Australia (three sites), and New Zealand. Portable calibration equipment, along with certified technicians, are

Table 8.1 Summary of FWD types by state highway agencies. After [45]

Quantities of FWDs Owned by State Highway Agencies, by Manufacturer		
Manufacturer	**Quantity of FWDs in Service (total)**	**Ages of FWDs (years, average)**
Dynatest	61	14
JILS	15	6
KUAB	6	14
Carl Bro	0	Not applicable
Other	0	Not applicable
Total	82	11

available from equipment manufacturers for onsite calibration. The certification of the calibration centers was originally operated by the Cornell Local Roads Program, but in 2010 the AASHTO Materials Reference Laboratory (AMRL) took responsibility for the program with oversight from the FHWA [50]. AMRL evaluates the centers and operators for compliance to AASHTO R32, Standard Recommended Practice for Calibrating the Load Cell and Deflection Sensors for a Falling Weight Deflectometer.

Back-calculation is generally used for the interpretation of the deflection data, with an objective of determining modulus values for each of the layers in the pavement. A generic structure for back-calculation is shown in Figure 8.1 [51]. Elastic layer theory is commonly used to evaluate deflection. Several computer codes are available for performing a back-calculation analysis. Both the SHRP LTPP program [52] and the Mechanistic Empirical Pavement Design Guide [53] selected the MODULUS program [54] for the analysis of flexible pavement deflections. This program has been updated periodically with the current version being Modulus 6.1.

8.2.2   Moving Measurement of Deflections

Operation of the FWD requires that traffic be disrupted and this has generally limited deflection analysis to project level applications [55]. Practical application of structural evaluation for network level analysis will require moving equipment that minimizes traffic disruption, preferably that travels near traffic speeds. While such equipment has been pursued for years, it is only recently with the advances in instrumentation and computers that

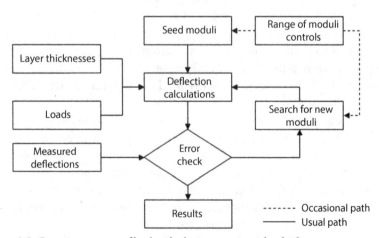

Figure 8.1 Generic structure of back-calculation process. After [51]

devices that can measure deflections continuously with minimal traffic disruption have become more feasible. The state of the technology for moving deflection measuring equipment was evaluated by Rada and Nazarian in 2011 [56]. In addition they assessed the best uses of the available devices and provided recommendations for further research. Five dynamic deflection devices were identified by [57]. These and others identified and summarized in Table 8.2 by Rada and Nazarian [56] were further evaluated. Based on a literature review and manufacture information, all devices were deemed to have some value for pavement engineering. However, the Quest Integrated – Dynatest HRWD and the Swedish National Road Administration RDT were dropped from consideration in the study as their continued development has been stopped or the prototype was not available.

Rada and Nazarian concluded that moving deflection devices were viable, but there are multiple issues that need to be resolved before this technology is ready for routine implementation. They identified several issues including: 1) need for further equipment development to increase speed of operation (in particular the Texas RDD), 2) repeatability and maintenance of details with special averaging, 3) measurement of deflection basins, 4) sensitivity of the deflection measurements to different pavement structures, and 5) identifying the minimum number of deflection points needed for specific applications.

The potential for using continuous deflection measurements for assessing pavement preservation and restoration needs was recognized by [55]. This study stresses that a moving deflection device must be capable of measuring deflection at intervals of approximately 1 ft. (300mm) or smaller, using loads similar to trucks (9,000 to 11,000 lbs [40 to 50 kN] per wheel or load assembly), and the device should be able to collect data without supporting traffic control measures. Sequential deflection measurements may be averaged to reduce noise in the measurements so the actual reporting distance between deflections may be greater than one foot. Based on a survey of state highway agencies network level the applications would include:

- Help identify "weak" (i.e., structurally deficient) areas that can then be investigated further at the project level;
- Provide network-level data to calculate a "structural health index" that can be incorporated into a PMS; and
- Differentiate sections that may be good candidates for preservation (good structural capacity) from those that would likely require a heavier treatment (showing structural deficiencies).

Table 8.2 Summary of moving pavement deflection device capability. After [56]

Device	Texas Rolling Dynamic Deflectometer (RDD)	Highway Rolling Weight Deflectometer (HRWD)	Rolling Wheel Deflectometer (RWD)	Rolling Deflection Tester (RDT)	Travel Speed Deflectograph (TSD)
Manufacturer	UT Austin	Dynatest Consulting and Quest Integrated	Applied Research Associates	Swedish National Road Administration and VTI	Greenwood Engineering UK Highway Agency
Estimated Cost	NIA	NIA	N/A	NIA	$2,400,000
Operational Speed	1 mph	20 mph	45 to 65 mph	60 mph	50 mph
Distance between Readings	2 to 3 ft	9 ft	0.5 in	0.001 s	0.80 in.
Applied Load	10 kips static + 5 kips dynamic	9 kips	18 kips	8 to 14 kips	11 kips
Deflection Sensor Accuracy	0.05 mils	NIA	± 2.75 mils	±10 mils	±4 mils
System Accuracy	NIA	1 mil at 6 mph	N/A	NIA	0.2 mils
Other Features	GPS Equipped	NIA	GPS Equipped	NIA	GPS Equipped
Number of Operators	2	NIA	2	2	2
Calibration Process	Yes	NIA	Yes	Yes	Yes

Initially, user agencies expressed interest in using moving deflection devices for project level evaluation; however, after follow-up interviews the states realized that current equipment limitations do not support the project level.

The RWD and the TSD were selected for further evaluation. The Texas RDD was not considered due to low operating speed. The study was conducted in two phases in which the technology evolved. The manufacturer upgraded the equipment between the two phases. In addition, a new device was introduced during the project that the researchers could not evaluate. With this in mind, the key issue is not the technology or specific findings of the research; rather, it is evaluation and implementation strategies for using new technology in an agency. The research team was not able to develop recommendations for selecting and using a continuous deflection measuring device. The current TSD device indicated adequate repeatability for network-level pavement management.

Although moving deflection equipment is in a developmental stage, several state highway agencies have evaluated potential applications. The RWD was developed by Applied Research Associates under several contracts with the FHWA. The RWD was evaluated in 13 states from 2003 to 2008. The results of these evaluations were documented in a series of reports by [58]. In addition, the Virginia Department of Transportation (VDOT) reported a RWD evaluation in 2010 [59]. Comparison of the RWD with FWD on two interstate sections found the results are not well correlated and the standard deviations of the RWD deflections were large and fluctuated with changes in surface type. The RWD was not recommended for pavements with low deflection values and uniform cross sections, i.e. interstate facilities. All these studies indicate that improvements are still needed to the RWD to make it effective even for network use.

In 2012 [60,61] reported on RWD evaluation in Louisiana. The authors stated this was the first project using the "updated" RWD. Data were collected on 16 sites with a wide range of pavement conditions. A comparison of RWD and FWD data found the mean center deflections were significantly different for 15 of the 16 sites. But the authors felt that both the RWD and FWD data gave reasonable results. This seems to be a strange finding. It was determined that the repeatability of the RWD was good. A Structural Number (SN) model was developed using the deflection measurements, and the SNs from the RWD and FWD showed good agreement. It was recommended that the use of the RWD be extended to other districts within the state.

8.2.3   Ground Penetrating Radar

To evaluate materials properties the thickness of each layer must be known, in addition to deflection and the type(s) of material in each pavement layer. Field cores have long been used for measuring thickness. However, this requires traffic control and has the same drawbacks as static deflection testing. In the mid 1990s ground penetrating radar (GPR) was being introduced as a technology for rapidly estimating pavement thicknesses. GPR technology includes both air-coupled and ground-coupled antennas. The air-coupled antennas operate above the pavement surface and can be operated at highway speeds; thus, they are suitable for network evaluations. Ground-coupled antennas are placed directly on the pavement, so they are limited to project level evaluation or situations where traffic control is provided, e.g. they can be used in conjunction with FWD data collection.

GPR capability for determining pavement layer thicknesses was evaluated by [43] with favorable results, using an air-coupled GPR operating at 10 to 20 mph. The GPR responses were calibrated by comparison with cores at the same locations. According to Maser, the study showed reasonable comparisons of +/- 8% when compared to thicknesses measured on cores.

An NCHRP Synthesis on the use of GPR to measure pavement thickness [44] found GPR to be promising, but at the time it had not matured to the point of routine application. Several state highway agencies subsequently investigated GPR with varying degrees of success and implementation [37,39–43]. The consensus of these findings is that GPR is reasonably accurate for determining the thickness of asphalt concrete layers, especially when calibrated against core measurements. The results for Portland Cement Concrete were not as favorable. The U.S. Corps of Engineers evaluated multiple technologies for measuring pavement thickness in 2011 [36] on 40 test locations including both asphalt and Portland concrete surfaces. The devices evaluated and the test results are summarized in Table 8.3. The MIRA (University of Minnesota) device uses ultrasonic technology for measuring pavement thicknesses. In addition to a GPR device, Olson Engineering provided an impact echo device and a multiple impact surface wave (MISW) device. All other devices used either air-coupled or ground coupled GPR technology. The researchers concluded that separate brands of devices are required for accurate evaluation of AC and PCC pavements. Air-coupled GPR worked well for AC pavements, but either seismic or ultrasonic devices performed better for PCC pavements.

Table 8.3 Summary of COE results for NDT thickness measurement devices. After [36]

Method	R^2 for All Test Locations[1]	Average Error[2] in.	% of Test Locations Measured[3]
MIRA (University of Minnesota)	0.98	0.51	70
Calibrated 1-GHz GSSI Horn Antenna (Infrasense, Inc.)	0.97	0.32	73
1-GHz GSSI Horn Antenna (Infrasense, Inc.)	0.96	0.51	73
Calibrated 1-GHz Pulse Radar Horn Antenna (ERDC)	0.95	0.40	63
Calibrated MISW (Olson Engineering)	0.91	1.15	78
MISW (Olson Engineering)	0.87	2.10	78
Calibrated 800-MHz, 1.2-GHz, 1.6-GHz, and 2.3-GHz ground-coupled antennas (MALÅ)	0.85	1.28	100
1-GHz Pulse Radar Horn Antenna (ERDC)	0.80	1.05	70
Calibrated Single 1.5-GHz Ground-Coupled GSSI Antenna (Infrasense, Inc.)	0.71	0.87	80
Single 1.5-GHz Ground-Coupled GSSI Antenna (Infrasense, Inc.)	0.68	1.10	83
Calibrated Impact Echo (Olson Engineering)	0.51	1.82	100
Calibrated 3D-Radar	0.41	2.11	100
Impact Echo (Olson Engineering)	0.39	2.29	90

Device	Coefficient of Determination[1]	Error[2]	Percentage[3]
Calibrated CMP Method with Two 1.5-GHz Ground-Coupled GSSI Antennas (Infrasense, Inc.)	0.38	1.37	80
800-MHz, 1.2-GHz, 1.6-GHz, and 2.3-GHz ground-coupled antennas (MALÅ)	0.38	2.37	100
Calibrated 2-GHz Ground-Coupled IDS Aladdin GPR (Olson Engineering)	0.34	2.35	100
CMP Method with Two 1.5-GHz Ground-Coupled GSSI Antennas (Infrasense, Inc.)	0.31	1.68	83
900-MHz and 1.5-GHz Ground-Coupled GSSI Antennas (Resource International)	0.22	7.19	100
Calibrated 900-MHz and 1.5-GHz Ground-Coupled GSSI Antennas (Resource International)	0.17	2.51	100
3D-Radar	0.03	3.88	100
2-GHz Ground-Coupled IDS Aladdin GPR (Olson Engineering)	0.03	4.48	100

[1] Coefficient of Determination calculated using the measured core thicknesses and the measured thickness from the device.

[2] Error was calculated as the absolute difference between the thickness measured with the devices and the actual thickness of the core. The errors for all the test locations were averaged for each method.

[3] Percentage of both AC and PCC test locations measured because not all of the test locations were measured for each device.

8.3   Destructive Structural Evaluation

Nondestructive pavement evaluation technologies have improved during the last 20 years, but there are still situations when it is beneficial to take core samples or to examine the layers by cutting a trench across the pavement. However, due to the expense and inconvenience to the motoring public, destructive evaluation is not a regular PMS function and is primarily used for research projects and forensic evaluations.

8.4   Structural Capacity Index Concepts

A structural capacity index should provide an estimate of either the maximum load a pavement can carry or how many axle load repetitions it can withstand. Some agencies definition of structural capacity includes pavement distress condition factors, e.g. the Kansas DOT Pavement Structural Evaluation (PSE) is "subjective and based on the condition of the pavement as indicated by the visual distresses and maintenance histories and the ability of the section to provide an adequate surface for the prevailing traffic" [62]. The preference herein is to call such an index a composite pavement condition index and reserve the concept of a structural capacity index as an indicator of the load carrying capacity of the pavement.

The concept of a structural capacity parameter was introduced as a Structural Adequacy Index (SAI) [3]. The key feature of SAI is the use of deflection measurements to rate the structural capacity of pavements. This concept has been applied in subsequent research for developing a structural capacity parameter.

Both empirical and mechanistic concepts were used to develop a Structural Strength Index (SSIF) [63]. Other authors have used SSI for a structural capacity index. To avoid confusion, the index developed by Scullion is referred to as SSIF since that is the dependent variable for Scullion's final equation. The SSI is determined from falling weight deflectometer data using the surface curvature index (SCI) and the deflection measured 72 inches from the FWD load. SCI is the difference between the deflections measured under the load and 12 inches from the center of the load. SSI values were established for three flexible pavement types: surface treatments, thin asphalt surfaces, and intermediate and thick asphalt pavements, as given in Table 8.4. Rigid pavements were not included in the study. The final SSI values from the tables were then corrected for rainfall and traffic, as:

$$SSIF = 100\left(SSI\right)^{1/(RF*TF)}$$

Table 8.4 Initial structural strength index values for TxDOT flexible pavements. After [63]

Surface Treated			Thin Asphalt			Intermediate and Thick Asphalt		
W7	SCI	SSI	W7	SCI	SSI	W7	SCI	SSI
<1.2	<20	1.00	<1.2	<15	1.00	<1.2	<10	1.00
	20–25.9	.80		15–20.9	.80		10–15.9	.80
	26–30.9	.60		21–25.9	.60		16–20.9	.60
	31–35.9	.40		26–30.9	.40		21–25.9	.40
	35–40	.30		31–35	.30		26–30	.30
	>40	.20		>35	.20		>30	.20
1.3–1.9	<20	.90	1.3–1.9	<15	.90	1.3–1.9	<10	.90
	20–25.9	.70		15–20.9	.70		10–15.9	.70
	26–30.9	.50		21–25.9	.50		16–20.9	.50
	31–35.9	.35		26–30.9	.35		21–25.9	.35
	36–40	.25		31–35	.25		26–30	.25
	>40	.15		>35	.15		>30	.15
>2.0	<20	.80	>2.0	<15	.80	>2.0	<10	.80
	20–25.9	.55		15–20.9	.55		10–15.9	.55
	26–30.9	.40		21–25.9	.40		16–20.9	.40
	31–35.9	.30		26–30.9	.30		21–25.9	.3
	36–40	.20		31–35	.20		26–30	.20
	>40	.10		>35	.10		>30	.10

where RF and TF are the rainfall and traffic factors given in Table 8.5. The traffic factors are based on the projected 20 year cumulative equivalent single loads in millions; the ESAL values are in the body of the table and the traffic factor is read from the top row, e.g. the traffic factor for a type 4 pavement with 25 million ESALs would be 0.85.

The SSIF approach is simple to understand and implement and can be scaled to provide an index that is compatible with other pavement condition parameters. The SSIF was added to the TxDOT pavement evaluation

Table 8.5 SSIF traffic and rainfall factors. After [63]

Pavement Type	Traffic Factors, TF					Rain Fall Factors	
	1.3	1.15	1.0	.85	.7	Inches/year	RF
4 HMAC>5.5"	<6	<11	<18	<26	>26	Inches/year	RF
5 HMAC 2.5–5.5"	<1.5	<3.1	<6.5	<21	>21	<20	1.0
6 HMAC <2.5	<0.5	<1.4	<2.7	<7.5	>7.5	21–40	.97
10 Surf. Treat.	<0.09	<0.24	<0.79	<3.4	>3.4	>40	.94

system, predecessor to the current pavement management information system, in 1987 [64]. However, an internal TxDOT study in 2000 found the SSI was not sensitive enough to effectively identify pavement sections with structural deficiencies. Accordingly, a method was developed based on the structural number (SN) as defined in AASHTO Guide for the Design of Pavement Structures, 1993, for a Structural Condition Index, (SCI) [64], to wit:

$$SCI = \frac{SN_{eff}}{SN_{req}}$$

where
  SCI = Structural Condition index
  SN_{eff} = the existing (estimated) Structural Number
  SN_{req} = the required Structural Number

The SN_{req} values were established using the AASHTO Guide for three levels of subgrade modulus and five levels of total 20 year accumulated traffic, as given in Table 8.6.

The resilient modulus of the subgrade, M_r in Table 8.6, is computed as:

$$M_r = 0.33 \times 0.24 \times \frac{P}{W_7 \times 72}$$

where
  P is the applied load, 9000 lbs
  W_7 is the deflection measured at 72 inches

Table 8.6 SN$_{req}$ for TxDOT structural capacity index. After [64]

Mr (psi)			20-Year Accumulated Traffic in ESALs				
			Very Low	Low	Medium	High	Very High
Category	Range	Average	50,000–945,000	945,000–1,687,000	1,687,000–2,430,000	2,430,000–3,172,000	3,172,000–50,000,000
			498,000	1,316,000	2,059,000	2,801,000	26,586,000
Low	1,000–5,400	3,200	4.3	5.1	5.3	5.6	7.1
Medium	5,400–7,500	6,400	3.5	3.9	4.2	4.3	6.0
High	7,500–40,000	24,000	2.3	2.6	2.8	2.8	3.9

The steps to compute SN_{eff} are [64]:

1. Normalize FWD deflections to 9,000 lbs.
2. Estimate the deflection, $W_{1.5Hp}$, at 1.5 times the total pavement thickness, H_p, using interpolation of the three sensors closest to the point at 1.5 times H_p.
3. Compute the Structural Index of a Pavement, SIP, from the deflection under the load, W_1, and $W_{1.5Hp}$, as:

$$SIP = W_1 - W_{1.5H_p}$$

4. Compute the SN_{eff} as:

$$SN_{eff} = k1 \times SIP^{k2} \times H_p^{k3}$$

where k1, k2, and k3 are given in Table 8.7.

Due to pavement variations, and consequently SCI variability, it was recommended that the number of deflection measurements be increased to two per half-mile section. Furthermore, the use of an average SCI is not suitable for selecting projects. The researchers thus recommended the use of two criteria [64]:

1. At least 50 percent of the (half-mile) sections that make up a potential project have a SCI value smaller than 1.0, and
2. A threshold percentage, to be calibrated by TxDOT, have a SCI below a minimum value. For example, if "20 percent or more of the sections are below a minimum SCI value of 0.70, then the section as a whole would be budgeted for MR activities."

Research for VDOT built on the TxDOT SCI concept created a Modified Structural Index (MSI), using the ratio of the effective and required

Table 8.7 Coefficients for determining SN_{eff} for SCI. After [64]

Surface Type	k1	k2	k3
Surface Seals	0.1165	−0.3248	0.8241
Asphalt Concrete	0.4728	−0.4810	0.7581

structural numbers [65]. The equations for computing the SN_{eff} and SN_{req} were modified to:

$$SN_{eff} = 0.4728 \times \left(D_0 - D_{1.5Hp}\right)^{-0.4810} \times Hp^{0.7581}$$

and

$$SN_{req} = a \times \left(log\left(ESAL\right) - 2.32 \times log\left(M_r\right) + \beta\right)^\gamma$$

The constants α, β, and γ depend on the road category as presented in Table 8.8.

MSI is a continuous variable and as such is not an index with defined bounds. The researchers introduced a threshold concept for the guidance or scoping of project type selection. An analysis of the VDOT pavement management and deflection data collected on Highway I-81 found that for MSI>1.0 pavement preservation treatments, e.g. nonstructural overlays, should be adequate. For MSI<1 a structural enhancement is needed. Thus, the project scoping decision is between structural overlay and reconstruction. The analysis of the Highway I-81 data indicated that a threshold value of 0.9 for MSI distinguished between the need for a structural overlay and reconstruction.

In another example, a structural strength index was developed for the Indiana Department of Transportation [66,67]. In this study, the FWD deflection measurement under the load was used to quantify the structural capacity. This measure was compared to functional indicators of pavement quality IRI rut depth, and to the INDOT pavement condition rating (PCR). Statistical tests indicated that the deflection was not correlated to any of the functional measures, indicating that structural condition was independent of functional condition. This should not be surprising since there is a time lag and today's structural index should only be a predictor of future performance. The structural strength index is computed as [66,67]:

Table 8.8 Road category constants used in the VDOT modified structural index. After [65]

Road Category	α	β	γ
Interstates	0.05716	9.07605	2.36777
Primary, divided or undivided	0.0600	8.89818	2.32629
High Volume Secondary	0.05919	8.77764	2.32729

Table 8.9 Coefficients of SSI model. After [66,67]

	Coefficients of Model		
Pavement Family	α	β	γ
Flexible Interstate	1.001	40.303	3.853
Flexible Non-Interstate NHS	1.004	66.811	3.106
Flexible Non-NHS	1.012	100.838	2.586
Rigid Interstate	1.035	14.301	3.056
Rigid Non-Interstate NHS	1.002	338.056	4.995
Rigid Non-NHS	1.072	23.6	1.999

$$SSI_{jk} = 100\left(1 - ae^{\frac{\beta}{(\delta_1)^\gamma}}\right)$$

Where

SSI_{jk} = Structural Strength Index for pavement family jk

α, β, γ = weighting constants determined through regression as given in Table 8.9

δ1 = deflection under the load, corrected for temperature

The family types are defined by pavement type: flexible or rigid; and highway classification: Interstate, Non-Interstate NHS, and Non-NHS.

The SSI rating of a pavement is based on a cumulative probability distribution curve of pavement deflections over the different pavement families, threshold deflections, and SSI values. Table 8.10 was developed for a qualitative assessment of structural rating.

8.5   Network versus Project Level Applications of Structural Capacity Evaluation

While deflection measurements are used for both network and project level evaluations, they are primarily a project-level tool. Until moving deflection devices reach an implementation stage, highway agencies must rely on static devices, primarily the FWD. If deflection measurements are used for network level pavement evaluation, it is with a much lower level of data intensity than used at the project level. Any data collection

Table 8.10 SSI thresholds for Indiana pavements. After [66,67]

Pavement	System	Measure	Excellent	Very Good	Good	Fair	Poor
Flexible	Interstate	Deflection (mil)	0–4	4–6	6–8	8–10	>10
		SSI	95–100	90–95	85–90	80–85	<80
	Non-Interstate NHS	Deflection (mil)	0–6	6–8	8–10	10–12	>12
		SSI	90–100	85–90	80–85	75–80	<75
	Non-NHS	Deflection (mil)	0–8	8–10	10–12	12–14	>14
		SSI	85–100	80–85	75–80	70–75	<70
Rigid	Interstate	Deflection (mil)	0–4	4–6	6–8	8–10	>10
		SSI	95–100	90–95	85–90	80–85	<80
	Non-Interstate NHS	Deflection (mil)	0–6	6–8	8–10	10–12	>12
		SSI	90–100	85–90	80–85	75–80	<75
	Non-NHS	Deflection (mil)	0–8	8–10	10–12	12–14	>14
		SSI	85–100	80–85	75–80	70–75	<70

program for project level evaluation should be designed or selected based on the anticipated variability of the pavement section being evaluated. For highway pavements, this may be at intervals of ten measurements per mile or six measurements per kilometer. Timing of the data collection should be coordinated with the design process such that the data are available in a timely manner for design. At the network level, both the intensity and timing of the deflection measurements must be determined. Based on an analysis of multiple years of FWD data, a measurement interval of three measurements per mile (approximately two measurements per kilometer) was recommended [62]. A measurement frequency of testing once every three years was recommended for network structural evaluation.

8.5.1   Staged Measurements

If an agency desires to consider deflection in a network analysis, say as part of project selection, it can be done by staging measurements as the British did on their motorways in the 1980s. First they measured their entire network for profile or serviceability and classified them as "very good, good, fair, poor, and very poor." Any pavement below the "good" category was surveyed for distress; others were passed over until the next round of roughness surveys. Those pavements below the "good" level in both serviceability (roughness and distress) then became candidates for deflection surveys. By this method, the amount of deflection testing was greatly reduced to no more than 20–25% of the motorway network. The deflections were made with a moving Benkelman beam traveling at walking speed in both wheel paths of the outer lanes with a sampling interval of 10–15 meters (about 20–40 feet). If anomalies were noted, for example a smooth (very good) pavement already suffering major distress, the deflection program also measured these sections. It seems that this quasi mixture of network with project-level PMS has been abandoned since the British Highway Agency is now seeking to implement a network-level PMS.

9

Evaluation of Pavement Surface Distress Condition Surveys

9.1 Purposes of Surface Distress Surveys

Pavement distress is an important component in defining the status of a pavement and can be useful in selecting appropriate preservation treatments. While deflection, roughness, and safety are also components of a pavement's condition, historically the term condition survey identified the process of evaluating surface distresses. Distress evaluation generally considers three factors: the type, severity, and extent of damage. Examples of distress types and descriptions are provided in many highway agency manuals as well as in the Long-Term Pavement Performance Distress Identification Manual [68].

9.2 Manual Methods for Distress Surveys

Manual condition surveys were the predominant method of data collection for many years. They were conducted either by walking along the pavement or through the windshield while driving, generally at slow speed. Only a few devices were available for automating the process. With

Table 9.1 Summary of agency pavement condition data collection. After [2]

	Method	Number of Agencies		
		Agency	Vendor	Total
Data Collection	Windshield	19	2	21
	Automated	23	21	44
Data Processing	Fully Automated	7	7	14
	Semi-Automated	16	14	30

advancements in sensors and computer technology, the state of the art in pavement condition surveys is rapidly advancing. Automated surveys are commonly used to capture an image of the pavement with either semi-automated or fully automated processing. Semi-automated processing has human interaction with the review of the pavement images, while fully automated relies on computer processing of the images to identify and quantify pavement distress. Each of these methods is used by highway agencies as shown in Table 9.1 [2]. About one third of the agencies rely on manual surveys, with most doing the survey in-house. About two thirds of the agencies use an automated method for collecting the data, with about an even split between collecting the data in-house versus using a vendor to collect the data. Of the agencies using automated methods, about two-thirds use semi-automated methods to reduce the data and one-third rely on fully automated data reduction.

The specific distress evaluation methods vary among agencies. Both ASTM and AASHTO have published standards for distress evaluation. The relevant AASHTO standards are:

- AASHTO R 36, Standard Practice for Evaluating Faulting of Concrete Pavements
- AASHTO R 48, Standard Practice for Determining Rut Depth in Pavements
- AASHTO R 55, Quantifying Cracks in Asphalt Pavement Surface
- AASHTO PP 67, Quantifying Cracks in Asphalt Pavement Surfaces from Collected Images Utilizing Automated Methods
- AASHTO PP 68, Collecting Images of Pavement Surfaces for Distress Detection

- AASHTO PP 69, Determining Pavement Deformation Parameters and Cross Slope from Collected Transverse Profile
- AASHTO PP 70, Collecting the Transverse Pavement Profile

9.3   Automated Survey Methods

As a practical matter, the technology used for measuring rutting is different from that used for cracking. Since rutting is distortions in the transverse pavement profile, the technology used to measure roughness (distortions in the longitudinal profile) can be applied and has matured further than the technology for fully automatic crack detection. The basic process for automatic crack survey technology is summarized in Figure 9.1 [69]. TRB Circular No. E-C156 is an excellent summary of the technology used for automated cracking surveys [70]. The technology for image capturing is developing rapidly. Analog (film) based systems have been supplanted with digital methods. The dominant sensor type used for capturing pavement images is the charged couple device (CCD) operating in either a line or area scan mode. The quality of the image is a function of the camera resolution, expressed in pixels, dynamic range, camera optics, and illumination. The other component of image acquisition is the compression of the image for storage. A 1-km section that is 4-m wide results in 1.024 GB of data at a dynamic range of 8-bit for 2,048 pixels per lane, and 4.1 GB for 4,096 pixels per lane. Compression is required to reduce the volume of the data to a manageable level.

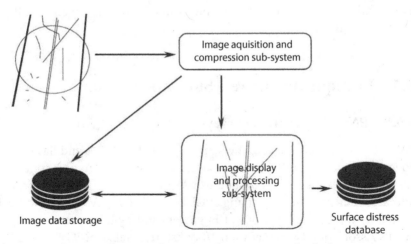

Figure 9.1 Schematic of automated pavement distress survey. After [69]

Laser illuminated technology was introduced in 2005 with the potential to improve the quality of images. Vendors are actively pursuing this technology as it overcomes many of the image quality issues associated with the artificial illumination needs of CCD cameras.

After the image data are stored, the key features of the images must be extracted to assess the pavement distress, using either a semi or fully automated process. A semi-automated process requires visual observation of the images to identify the distresses. This involves a human observer examining the images to identify and facilitate the process. Vendors have developed proprietary systems for this task. The semi-automated process has some intuitive appeal as the image recognition process is easier for humans than machines. However, this process can induce bias and errors.

9.4   Types of Distress

A detailed national standard was prepared for evaluating pavement distress for the Long-Term Pavement Performance Program [71]. The distress types identified in this manual are listed in Table 9.2. The manual provides written descriptions, photographs, and measurement methods for each distress type and severity. While this is a good resource for identifying pavement distress types and is available on the internet, it is a research manual and is not a practical field manual. Figure 9.2 [1] summarizes the percent of state DOTs that collect data on the various distress types. More specific information on distresses and definitions used by individual agencies must be obtained directly from the agency or PMS software venders.

9.5   Examples of Distress Survey Procedures

9.5.1   PAVER™ Distress Surveys

PAVER™ is a system that was developed in the late 1970s and has undergone a number of updates and modifications since then. The condition survey method they use is unique to PAVER™. They combine distresses into a pavement condition rating (PCR). Currently PAVER™ 6.5 is supported by the U.S. Army Corp of Engineers and by the American Public Works Association (www2.apwa.net/bookstore/detail.asp?PC=PB.APAV).

Table 9.2 Distress types defined in SHRP distress survey manual. After [71]

Asphalt Concrete Surfaces	Jointed Portland Cement Concrete Surfaces	Continuously Reinforced Concrete Surfaces
A. Cracking	**A. Cracking**	**A. Cracking**
1. Fatigue Cracking	1. Corner Breaks	1. Durability Cracking ("D" Cracking)
2. Block Cracking	2. Durability Cracking ("D"Cracking)	2. Longitudinal Cracking
3. Edge Cracking	3. Longitudinal Cracking	3. Transverse Cracking
4. Longitudinal Cracking	4. Transverse Cracking	**B. Surface Defects**
5. Reflection Cracking at Joints	**B. Joint Deficiencies**	4. Map Cracking and Scaling
6. Transverse Cracking	5. Joint Seal Damage	4a. Map Cracking
B. Patching and Potholes	5a. Transverse Joint Seal Damage	4b. Scaling
7. Patch Deterioration	5b. Longitudinal Joint Seal Damage	5. Polished Aggregate
8. Potholes	6. Spalling of Longitudinal Joints	6. Popouts
C. Surface Deformation	7. Spalling of Transverse Joints	**C. Miscellaneous Distresses**
9. Rutting	**C. Surface Defects**	7. Blowups
10. Shoving	8. Map Cracking and Scaling	8. Transverse Construction Joint Deterioration
D. Surface Defects	8a. Map Cracking	9. Lane-to-Shoulder Dropoff
11. Bleeding	8b. Scaling	10. Lane-to-Shoulder Separation
12. Polished Aggregate	9. Polished Aggregate	11. Patch/Patch Deterioration
13. Raveling	10. Popouts	12. Punchouts
E. Miscellaneous Distresses	**D. Miscellaneous Distresses**	13. Spalling of Longitudinal Joints
14. Lane-to-Shoulder Dropoff	11. Blowups	14. Water Bleeding and Pumping
15. Water Bleeding and Pumping	12. Faulting of Transverse Joints and Cracks	15. Longitudinal Joint Seal Damage
	13. Lane-to-Shoulder Dropoff	
	14. Lane-to-Shoulder Separation	
	15. Patch/Patch Deterioration	
	16. Water Bleeding and Pumping	

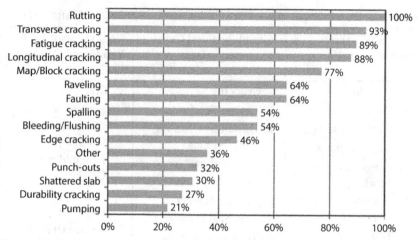

Figure 9.2 Distress types measured by state highway agencies. After [1]

9.5.2   FHWA Network Distress Collection Protocols

In 1996 the FHWA realized the need to have uniform, simple distress data collection protocols and funded a project with Texas Research and Development Foundation [72]. To achieve uniformity, new and/or modified protocols were developed considering available technology, practices within larger state DOTs and compromises necessary to facilitate network-level data collection by modern and emerging automated data collection equipment technology.

The primary innovation of this project was to extend the LTPP technology for effective network PMS data collection. First, the data collection should give enough detail and accuracy to allow PMS decisions at the network-wide level. Second, the data should be collected in low cost rapid condition surveys. Third, the protocols set a standard baseline for technology development and improvement in fully automated distress measurement equipment.

The protocols for faulting, roughness, and rut depth all are strictly oriented towards automated data collection at the network level. The cracking protocols were more complicated relative to automation but were simplified with vendor input to provide better uniformity.

Six separate protocols were developed including three automation-oriented protocols for pavement condition factors: 1) roughness, 2) rut depth, and 3) faulting. Three cracking protocols were developed for automated data collection for 1) asphalt surface pavements, 2) continuously reinforced concrete, and 3) jointed concrete pavement. Distresses collected by either

Table 9.3 Five sections with consistent characteristics. After [72]

1. Purpose	1a. Applies to National Highway System
	1b. Network Level Survey
2. Scope	2a. Specifications for equipment or instruments are not included.
	2b. Briefly describes what distress is surveyed.
3. Measurement	Defines the distresses to be measured
4. Recording and Aggregation of Data	4a. Defines severity levels
	4b. Describes how data is to be recorded
	4c. Data collection section length is between 0.15 km and 1.0 km. A constant section length is recommended. (*)
5. Quality Assurance	5a. Operator training is recommended.
	5b. Daily equipment quality assurance procedures are recommended.
	5c. Periodic equipment quality assurance procedures are recommended.

*NOTE: A data collection section differs from a pavement management section. A data collection section is a length of road over which one survey is performed. A pavement management section is defined by the agency as a length of road subject to the same rehabilitation strategy. It consists of from one to a combination of several data collection sections. The protocols address data collection sections. Pavement management section definition and data aggregation are the responsibility of each agency using a PMS.

manual or automated methods are similar within a particular pavement type protocol (AC, JCP, or CRCP). Each protocol consists of five sections with consistent characteristics across the protocols as shown in Table 9.3.

Roughness Protocol

Great attention was given to a standard roughness protocol, a "main stream" methodology for highway agency standardization. After a review of available alternatives, the international roughness index was selected [73].

Rutting

For routine automated data collection at the network level, as shown in Figure 9.3, a three-point rut depth measurement was recommended. This gives a stable indicator of rutting in the data collection section and is highly correlated with complex rutting data obtained by five-point, seven-point, or nine-point transverse measurements.

Rut Depth Measurement

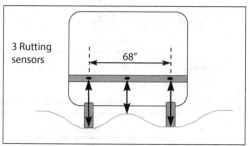

Figure 9.3 Three-point rut depth. After [72]

Faulting

Faulting is a relatively straightforward measurement and keyed to the measurement using the Georgia Fault Meter [68]. Automated methods for faulting are available from vendors such as Surface Dynamics, Inc. [74] and by International Cybernetics [75]. The faulting is summarized by distribution of fault severity.

9.5.3   Cracking Measurements

Perhaps the most innovative thinking put forward in the protocols in [72] is for cracking. Typically windshield type surveys or walking surveys were used for network-level cracking measurements. Other methods include the small but intensive sample taken for PAVER. Clearly, load associated cracking is found most often in the wheel paths, and the outside wheel path is a better indicator of load associated cracking than the inside wheel path.

Cracking in [72] was limited to three types: 1) environmental, or transverse cracking, 2) load, such as fatigue or wheel path cracking, and 3) miscellaneous, including longitudinal cracking not in the wheel path, block cracking not in the wheel path, and random or diagonal cracking. These all trigger the same type of maintenance.

These protocols were considered by AASHTO, but it appears there was a consensus that they were too simplistic in their PMG-2, Pavement Management Guide [76].

9.6   Equipment for Distress Evaluation

Of the various types of distress collection equipment used in 1994 [3], only one survives and all its components have been changed, once, twice, three times or more in the last 20 years. This illustrates the volatility of this part of PMS. There have been many comparisons of automated equipment since 1994. Example descriptions of two of these, Fugro Roadware ARAN and Pathway Services Inc., are discussed herein. Other descriptions can be obtained in the vendor's website or through direct contact, and this extends to other countries in Europe, Australia, New Zealand, and elsewhere.

The following five sections are updates on vendors and methods but not intended as endorsements of any particular one.

9.6.1   Comparison of Vendor Performance

A comparison of the performance of three vendors using semi-automated data collection to "reference" data collected with a manual survey is described in [77]. Figure 9.4 shows that all of the vendors underestimated the fatigue cracking. The data for the individual distresses were compared to the pavement condition rating (PCR) used by NCDOT. PCR is a composite condition statistic that combines fatigue cracking, transverse cracking, rutting, bleeding, patching, and oxidation. Figure 9.5 compares the pavement condition rating obtained by the automated equipment to results obtained with manual surveys collected by several raters. The line of equality was based on the average of the raters, and the vertical bars indicate the

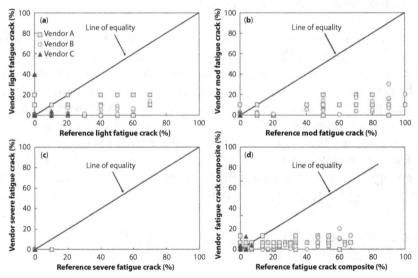

Figure 9.4 Comparison of fatigue cracking measures to three semi-automatic vendors and manual surveys. After [77]

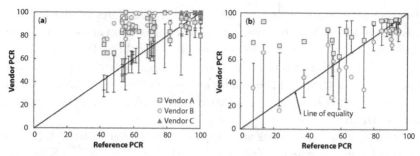

Figure 9.5 Comparison of pavement condition rating of three semi-automatic vendors and manual surveys. After [77]

range of PCR results from the manual surveys. This figure demonstrates that manual ratings suffer variability problems. The vendors consistently overestimated PCR for flexible pavements.

Although fully automated data collection has been in development since the late 1980s, the problem is still challenging. As stated in [70]:

> After many decades of research and development, the industry and developers as a whole have been frustrated with the slow pace of progress in fully automated systems for pavement condition survey. Even though the 2-D laser imaging introduced in 2006 provides shadow-free 1-mm images, the difficulty in providing pavement engineers fully automated results still remains.

Technology may be coming available that can make reliable fully automated pavement condition assessment a reality. Recently 3-D imaging technology has been adapted to the pavement condition survey problem. The potential of producing 3-D images with resolution of 1 mm or better in all three dimensions with associated data reduction software has been demonstrated [78-80]. With the rapid development of automated survey technology, there is a need for national coordination of the effort for collecting consistent and meaningful data. A pooled fund study is being developed to provide the type of support to this research area similar to the effort for improving profile measurements [81].

9.6.2   Synthesis of Pavement Distress Collection Techniques, 2004

It is useful to agencies making decisions about distress collection methods to know what other agencies are doing. Such a synthesis was carried out by NCHRP [82]. To expedite the gathering of information, a question-naire was sent to all U.S. state agencies, Canadian provinces, the World Road Association, and FHWA. A total of 56 responses were received from 43 state DOTs, two FHWA offices, 10 Canadian provinces, and Transport Canada. Other information was acquired through a literature search and more than 150 references are given in the synthesis report. The study found that essentially all North American highway agencies are collecting pave-ment condition data with automated means, usually with integrated vehi-cles. Fifty-two of the agencies collect roughness data, 50 agencies collect rut depth measurements, and joint faulting measurements are taken by 30 agencies as are pavement surface images. Automated processing of surface distress is used by only 14 of those agencies. The others use manual data reduction techniques. Thirty-three of those agencies used vendors to col-lect at least some of the automated data. In some of the agencies, the vendor collected sensor data but the agency continued to collect surface distress with manual surveys. Eighteen of the agencies procured their automated services through a Request for Proposals. Seven agencies used a Request for Qualification approach and eight used advertised contract and low bid approach.

A typical contract is for two years although some agencies use one year and one used a four-year period. Most of the agencies provided for negoti-ated contract extensions. Twenty-two of the agencies had QA provisions in their contracts and 12 had price adjustment clauses. It is interesting to look at pricing from 1994. IRI information ranged from a low cost of $1 per mile up to $170 per mile for vendor collection and processing of images and sensor data in an urban environment. About $50 per mile was the approximate average for that total work. It should be remembered, however, that in urban locations and extremely remote locations, prices can vary widely. Some agencies had done extensive QA development, and the Canadian provinces were exceptionally vigilant in that area and had established procedures that could be used for an international approach. It might be noted that most agencies do not build and operate equipment in-house.

Not all agencies are enamored with automatic procedures because they believe that data quality is compromised. On the other hand, several felt that data quality improved through automation. The difference may lie in the QA procedures involved. The synthesis noted that there is a difference between the state-of-the-art (what can be done) and the state of the prac-tice (what is being done) by certain vendors. Vendors face updating their equipment on a regular basis, but they can only do that if they obtain an adequate return on their investment.

What follows are a few selected statistics from [82], which the reader may find useful. The synthesis spent some time discussing Data Monitoring Frequency, Data Reporting Interval, and Linear Referencing, items not deeply discussed in this book but which are important. Based on the aforementioned survey, Table 9.4 shows the number of agencies using automated collection, automated processing, type of image capture, and the protocols they are using if any. They also show as in-agency or by contract and whether they collect data items such as cracking, IRI, rutting, and joint faulting. It is interesting that only nine of the agencies report using an identifiable cracking protocol, and five of these reported using LTPP, which is patently impossible by automated methods since LTPP collects more than 30 types of distress. Thirty-one used an AASHTO or ASTM IRI protocol, and 38 used their own rutting protocol. Joint faulting measurements were much less popular among the agencies.

Table 9.4 Overview of agency pavement data collection and processing (Number of Agencies) After [82]

Activity	Entity/Process	Data Item			
		Cracking	**IRI**	**Rutting**	**Joint Faulting**
Automated Collection	Agency	10	31	30	21
	Contract	20	23	21	12
Automated Processing	Agency	7	—	—	—
	Contract	7	—	—	—
Image Capture	Analog	16	—	—	—
	Digital	17	—	—	—
Sensor Data Collection	Laser	—	44	30	23*
	Acoustic	—	3	15	—
	Infrared	—	4	2	—
Protocol	AASHTO	4	12	6	4
Use	ASTM	—	19	—	—
	LTPP	5	—	—	—
	Other	21	16	38	10

*By sensor.
Notes: LTPP = Long-Term Pavement Performance; IRI= International Roughness Index.

Table 9.5 shows the frequency with which the agencies collect data. The predominant frequency was two years, but 26 data agencies collect smoothness/roughness data on a one year frequency along with rut depth.

The distance interval over which automated measurements were reported by agencies is summarized in [82]. As might be expected, there was a wide divergence among the agencies.

Table 9.6 shows a summary of the linear reference methods used in automated monitoring. While there is a lot of discussion of going to GPS Linear Referencing Systems or other improved linear referencing systems, as of 2004 the majority of the agencies continued to use mile point/post. Eight to 15 of the agencies were using latitude and longitude from GPS; the others used link-node and log mile.

Table 9.7 shows the type of equipment in use by states and provinces in 2004. Eight types of equipment are reported with one of those "agency

Table 9.5 Summary of automated monitoring frequencies employed (Number of Agencies). After [82]

Frequency (years)	Cracking, etc.	Smoothness/ Roughness	Rut Depth	Joint Faulting
1	9	26	24	10
2	18	20	20	13
3	2	4	4	0
Other	1	2	2	0
Total	30	52	50	23

Table 9.6 Summary of linear-reference methods using in automated monitoring (Number of Agencies). After [82]

Method	Cracking, etc.	Smoothness/ Roughness	Rut Depth	Joint Faulting
Mile Point (post)	33	46	35	23
Latitude-Longitude	12	15	14	8
Link-Node	5	5	5	2
Log Mile	3	1	1	0
Other	2	1	1	0

Table 9.7 Equipment in use. After [82]

Supplier	Agencies Using
Dynatest and Law	5
GIE Technologies	2
International Cybernetics Corporation (ICC)	9
INO	2
Pasco/CGH/ERES	1
Pathway Services	9
Roadware Group, Inc.	19
Agency Manufactured	1

manufactured," which was probably Texas DOT. Fugro-Roadware-ARAN, called "Roadware Group Inc.," was the predominant agency with 19; Pathway Services and ICC tied with nine each.

Finally, the report includes three case studies for state DOTs: Maryland, Louisiana, and Mississippi. Unfortunately, they do not give case studies for large active departments like North Carolina, Texas, and California. Table 9.8 for Louisiana sensor data collection requirements is worthy of note since it gives a full report of details of data collection in the Louisiana highway agency.

A word of caution on how notoriously difficult it is to get busy people from state agencies to fill out a detailed questionnaire. For this reason, no one can be 100% sure of the numbers presented, whether or not any of these agencies changed the year before the survey, the year or the decade after the survey. Nevertheless, the tables do provide a reasonable estimate of what goes on.

9.7 Summary of Pavement Distress Scores Used by State DOTs

In 2009 a study was carried out in Texas [83] with typical survey techniques although many of the agencies did not respond. Twenty-eight state DOTs and the District of Columbia reported using visual inspection by raters for data collection; eight states used automated or semi-automated methods (Alabama, Iowa, Louisiana, Maine, Oklahoma, South Carolina,

Table 9.8 Louisiana sensor data collection requirements. After [84]

Variables	Roughness	Rut Depth	Faulting
Scope	All pavements	Asphalt surfaces	Jointed concrete
Definition	Longitudinal profile, both wheel paths	Rutting of each wheel path	Elevation difference across joint (trailing slab lower)
Sampling	Max., 0.3 m (1 ft)	Max., 3 m (10 ft)	All transverse joints
Calculation and Statistics	IRI, each wheel path and average of both wheel paths	Each transverse profile of both wheel paths, for section report average	Wheel path absolute elevation difference averaged for each joint, for section report average
Units	Inches/mile	Inches [nearest 2.5 mm (0.1 in.)]	Inches [nearest 2.5 mm (0.04 in.)]
Equipment Configuration	Lasers and accelerometers, both wheel paths	Min., 3 laser sensors	Lasers in right wheel path
Standards	ASTM E950, HPMS Field Manual Class II	None given	None given
Precision and Bias	Max. error of 5% bias or 0.3 m/km (20 in/mi) whichever is less	Contractor to provide	Contractor to provide
Initial Verification	Section comparison of longitudinal profile with Class I profiling instrument and LADOTD's SD laser profiler	Test section comparison with field measurements provided by LADOTD	Test section comparison with field measurements provided by LADOTD
Ongoing Quality Monitoring	QA/QC sections	QA/QC sections	QA/QC sections
Special Requirements	Correct/report low-speed sections; capability of monitoring data collection in real time in the data collection vehicle	Capability of monitoring data collection in real time in the data collection vehicle	Capability of monitoring data collection in real time in the data collection vehicle
Reporting Frequency	0.16 km (0.10 mi)	0.16 km (0.10 mi)	0.16 km (0.10 mi)

Vermont, and Virginia); seven states had raters evaluate pavements using images or video logs (Connecticut, Illinois, Michigan, Missouri, Nebraska, Pennsylvania, and Tennessee). One state, Arkansas, used only the international roughness index as a basis for rating pavements. It is unclear what the remaining states used.

The following definitions were used in this study.

> *Type*—distress or condition categories (e.g., shallow rutting, deep rutting, longitudinal cracking, transverse cracking, alligator cracking, ride, etc.).
>
> *Extent*—the amount of distress present on the pavement section being rated (e.g., for alligator cracking, TxDOT uses percent of wheel path in a 0.5-mile section to measure the extent of alligator cracking distress.
>
> *Severity*—the degree of distress (e.g., rutting can be measured at 50 percent of wheel path [extent], but severity is addressed by measuring the depth of rut).
>
> *Distress*—For TxDOT, distress means cracking and rutting and does not include ride; distress scores for each type of distress are combined to determine a distress score. This may not be the case for other states.
>
> *Condition*—For TxDOT, pavement condition is the combination of distress scores and ride. Again, this may not be the case for other states.

TxDOT used extent and type of distress. Severity level was only considered for rut depth and ride quality. Twenty-nine states used extent and severity of distresses. Seven states used extent and type of distresses (California, Illinois, Iowa, Oregon, Pennsylvania, South Dakota, and Wisconsin). For the remaining 13 states and the District of Columbia, it was unclear what was used in terms of measured attributes.

9.7.1 Rating Scales and Levels of Acceptability

TxDOT reported a 5-level scale for scoring condition, distress, and ride score: very good, good, fair, poor, and very poor. These may well have been derived from the present serviceability index scale of 0 to 5 defined at the AASHO Road Test. It was interesting to note that 11 other states used a 5-level scale, seven states used a 4-level scale, and eight states used a 3-level scale. Ten states and the District of Columbia used other scales. There was no information for the remaining 13 states in terms of distress alone.

Table 9.9 shows the distress acceptance levels on a 100 point scale for 11 states [83]. It is important to note that the acceptability level ranges from a low of 40 in New Hampshire to a high of 75.1 in Oregon. This unfortunately makes it complicated to compare results among all 50 states. It also illustrates the need to reconsider the present serviceability rating/present serviceability index [15]. Such a scale provides uniformity and consistency.

Table 9.10 shows levels of acceptability on a 5-point scale for 10 agencies [83]. Some of the state agencies are listed on both tables as using both a 5-point and a 100-point scale. Again, there was little uniformity in the level used. These inverted scales make no sense to the riding public. Moreover, it is difficult for legislators to comprehend the differences. Adoption of the present serviceability index PSI scale could solve this problem and return such reporting back to the quality that existed in the 1960s-1970s after the AASHO Road Test.

9.8   Example Equipment: Fugro, Roadware-ARAN

Roadware has been active in pavement data collection for over three decades. They provide both equipment sales and data collection services. In the 2000–2010 timeframe, they were acquired by Fugro and now provide both equipment sales and services under the Fugro Roadware brand,

Table 9.9 Distress acceptance levels on 100-point scale. After [83]

Georgia	75–100 is good to excellent
Iowa	60–80 is good, 80–100 is excellent
Montana	63–100 is good
Nebraska	70–89 is good; 90–100 is very good
New Hampshire	40–100 is acceptable
North Carolina	Greater than 80 is good
Ohio	75–90 is good; 90–100 is very good
Oregon	75.1–98 is good; 98.1–100 is very good for NHS
Vermont	40–100 is acceptable
Virginia	70–89 is good; greater is excellent
Washington	50–100 is good

Table 9.10 Levels of acceptability on a 5-point scale. After [83]

California	2 is good; 1 is excellent
Delaware	3–4 is good; 4–5 is very good
Idaho	3–5 is good
Kentucky	3.5–5 is good
Michigan	1.0–2.5 is good
New Mexico	Greater than 3 is good for Interstate Highways; greater than 2.5 is good for all other highways
Oregon	2.0–2.9 is good; 1.0–1.9 is very good for non-NHS
South Carolina	3.4–4.0 is good; 4.1–5.0 is very good
Tennessee	3.5–4.0 is good; 4.0–5 is very good
West Virginia	4 is good; 5 is excellent

Figure 9.6 ARAN 9000. After [85]

using the ARAN (Automatic Road Analyze) brand name. The following provides a description based on information in [85].

The ARAN 9000 (Figure 9.6) is one of the most complete roadway data collection vans available in the world. It incorporates many capabilities and options, depending on budget and price. According to Fugro, the ARAN's robust platform can deliver the following:

- 50% reduction in computing hardware over the previous platform with the same functionality
- Data base driven systems

- Robust, fault tolerant systems
- Plug and play system integration
- Microsoft.net platform
- Real-time sub-cm data synchronization
- Advanced mission management software
- Increased portability of subsystem components
- Global solution with interfaces in several languages
- User friendly operating system to minimize training costs and operator error
- Industry-defining warranty
- Dynamic architecture supporting future upgrades.

Fugro Roadware also manufacturers the ARAN 8000 and the ARAN 7000, as described in [85].

Among Fugro's tools to aid in collecting data with good tolerances for several linear referencing systems (LRS), are identification features such as district, county, route, direction, lane, or other identifiers important to the client. Vehicle distance measurement (DMI) is provided and is synchronized with all other ARAN data collected, such as GPS. ARAN has a roughness (IRI) system based on the Fugro Roadware's Laser SDP/2 (South Dakota Profiler) which is a Class 1 profiling device under ASTM E950 and proven over a range of agency requirements including AASHTO or 56-10 certification. Additional technical specifications for the roughness system can be obtained from Fugro Roadware. Fugro also provides a Pave3D (LCMS) three-dimensional model of the road surface for automated crack detection.

Roadware also provides optional information on pavement texture obtained from Fugro Pave3D system and purports to provide GPR data at traffic speed (ground penetrating radar) for estimating pavement thickness. This is an important characteristic, but information that validates the accuracy of these GPR data taken at traffic speed is not available. It seems clear that Fugro Roadware ARAN is a widely used data collection vehicle. A listing of clients is not provided herein, but is expected to be available from Fugro Roadware if desired.

9.9   Example Equipment: Service Provider-Pathway Services Inc.

Pathway Services Inc., located in Tulsa, Oklahoma, has been providing and developing full-service PathRunner Data Collection Vehicles and data collection services to transportation agencies from the early 1990s. They also have offices in Central America. The information provided here is obtained from their website, www.pathwayservices.com [86]. Pathway Services reckons to have measured more than two million center-line miles of highways in the United States. They currently list 20 DOT clients, including the Mexican Federal DOT.

The Pathway Services Vehicle (Figure 9.7) provides information from an inertial road profiler, video logging, asset inventory, individual sensor data, and distress data including faulting calculations. They also provide PathView 2, a desktop data viewer for management and reporting software, as well as PathWeb, a technology that allows all data and images for a particular location to be viewed anywhere, with an internet connection tied to satellite imaging of Google Maps, Bing Maps, ESRI, and state collected orthogonal imaging.

The heart of the PathRunner XP data collection vehicle is an inertial road profiler (ASTM E950 specification, Class 1 requirements) which collects the longitudinal profile in both wheel paths and can calculate IRI with either the quarter car simulator or the half car simulator. The PathRunner XP road profiler has been certified at the Texas Transportation Institute, Minnesota Department of Transportation, and Washington State Department of Transportation test sites. The vehicle contains user-friendly graphical interface displays, where real-time graphs allow the operator to verify the proper operation of sensors and imaging equipment before, during, and after data collection. Real time data processing display and logic

Figure 9.7 PathRunner XP Collection Vehicle. After [86]

analysis ensure that data falls within valid ranges and image thumbnail views are provided to give real time quality capability.

Curvature data is collected by gyroscopes mounted inside the vehicle. These gyroscopes have a 200th of a degree resolution and are sampled at fixed distances approximately every five feet. This data is synchronized with the images and location information and/or filtered to calculate the beginning of a curve or point of curvature, the end of curve, the k value of the curve, and the length of the curve.

As a standard service, both cracking and patching data can be reported using the collected imaging and crack detection software. It automatically measures the length of the distress feature and creates a distress data base synchronized with the sensor data and video images. The system can be programmed for calculation of various distress indices, such as the PAVER PCI system or a client's specific defined system.

The company provides PathView 2, a desktop software application that integrates all of the pavement surface sensor data, digital images, and location data into a single interface on screen. The software provides easy import/export to other applications, with built-in functions such as dynamic find features, zooming within an image, photo surface distress measurement, and image brightness controls.

PathWeb allows all data and images to be reviewed from internet connection and leverages the satellite imaging of Google Maps, Bing Maps, ESRI, and state collected orthogonal maps. The user-friendly interface allows users to access full resolution imaging and plot GIS with a click of the mouse. Rutting and roughness, for example, can be colored coded by severity to provide statewide condition maps quickly.

9.10   Application of Distress Data

Distress is the precursor (leading indicator) for serviceability (smoothness) and performance. It is also a key indicator of the type, extent, and timing of maintenance and/or rehabilitation intervention needed. In other words, distress is the "what and when and where" for annual and long-term budgeting and programming decisions.

10

Evaluation of Pavement Safety

In the past several decades, there has been a huge growth and awareness in the importance of road safety as a public health issue. Prior to 1994, federally mandated highway regulations only mentioned skid resistance as part of the provision of safe, smooth roads and, in particular, as preventive reduction in wet-pavement crashes. But little guidance was given on how to achieve good skid resistance. With the inclusion of pavement surface characteristics in the Strategic Highway Research Program (SHRP) Long Term Pavement Performance (LTPP) study came an increase in research into the role and importance of pavement surface texture as a means to improve skid resistance and crash reduction. Work done under PIARC [87] led to the development of the International Friction Index (IFI) in 1995 and subsequent development of a friction management system framework in 2009 [88]. This describes the key components of pavement safety focusing on pavement surface texture and measurement tools.

Excellent resources are now available to designers and practitioners on pavement friction mechanisms, surface characteristics, materials (aggregate engineering, in particular), measurement, design and specifications from Federal Highway Administration (FHWA) circulars and manuals, Transportation Research Board (TRB) journals and conferences, American

Association of State Highway and Transportation Officials (AASHTO) manuals and guides, and international guides (Canada, United Kingdom, and Australia amongst others). The following references are key documents to be consulted [87–89].

10.1   Major Safety Components

The components listed in [3] are still valid and research continues to link them robustly to crash reduction. While road safety research includes study of the confluence—road characteristics, driver behavior, and vehicle characteristics. Titus-Glover [90] further defined vehicle collision factors directly attributed to the road in Table 10.1.

10.2   Skid Resistance Evaluation

Skid resistance is a complex interaction between the tire and pavement surface, as seen in Table 10.1, and is directly related to pavement surface

Table 10.1 Road-specific collision contributing factors. After [90]

Factor Affecting Collision	Description
Highway Design	Pavement width, alignment, curves, side-slope, terrain, number of access points, and intersections/interchanges
Pavement Characteristics	Micro-texture, macro-texture, mega-texture (unevenness), lateral and side-force friction, material properties, rut depth, and temperature
Traffic	Volumes, speed, congestion, percent trucks, and work-zones/construction
Vehicle Operating Parameters	Slip speed, braking action, and driving maneuvers (turning and overtaking)
Tire Properties	Footprint, tread design and condition, rubber composition and hardness, inflation pressure, and load
Environment	Temperature, climate, wind, water (precipitation form and density), contaminants (snow, ice, anti-skid material, mud, debris, and oil or other spills), visual distraction such as glare, nighttime driving conditions, etc.

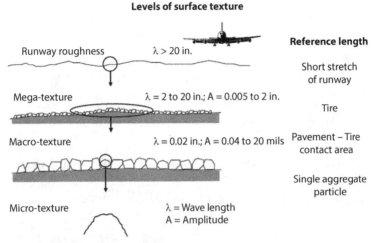

Levels of surface texture

Runway roughness — $\lambda > 20$ in. — **Reference length**

Short stretch of runway

Mega-texture — $\lambda = 2$ to 20 in.; $A = 0.005$ to 2 in.

Tire

Macro-texture — $\lambda = 0.02$ in.; $A = 0.04$ to 20 mils — Pavement – Tire contact area

Single aggregate particle

Micro-texture — $\lambda =$ Wave length $A =$ Amplitude

Figure 10.1 Pavement texture categories. After [91]

texture which is categorized into mega (large scale irregularities that contribute to pavement roughness), macro, and micro texture, as illustrated in Figure 10.1 [91].

10.3   Basic Concepts of Skid Resistance and the Importance of Pavement Texture

Micro-texture is fine-scale roughness of the aggregate which provides adhesion at low speed. In Asphalt Concrete pavements micro-texture is provided by the aggregate surface, and in Portland Cement Concrete (PCC) pavements, it is provided by the fine aggregate in the mortar. It is largely a function of aggregate shape characteristics and mineralogy (clean, strong, hard, and durable aggregate will resist polishing under tire wear) and is particularly important at reducing loss of skid resistance when the surface is wet. The dividing line between micro and macro texture is generally considered to be 0.5mm. Macro-texture is larger scale than micro-texture and provides 90% of the friction at speeds above 90km/hr. Macro-texture is measureable, as shown in Figure 10.2, and is a function of the mix properties, compaction method, and aggregate gradation in Asphalt Concrete pavements, or texturizing in Portland Cement Concrete pavements. Surface texturizing in PCC pavements is part of the design process and should be selected carefully: in both pavement types, aggregate petrography and engineering is important.

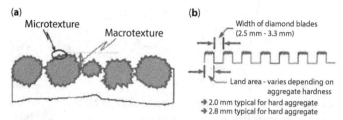

Figure 10.2 Macro v micro texture for a) asphalt concrete and b) portland cement concrete pavements. After [91]

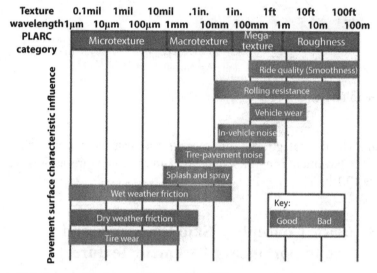

Figure 10.3 Pavement texture continuum. After [87]

In addition to providing friction benefits, macro-texture will reduce splash and spray as well as headlight glare, all of which contribute to improved visibility and safer driving. However, increased macro-texture also contributes to higher noise levels, tire-wear, and fuel consumption, particularly in the case of tined PCC pavements [92]. The continuum of surface texture from pavement roughness to micro-texture and how they each contribute to ride quality, vehicle operating costs, noise, and tire wear is further illustrated in Figure 10.3 [87].

Due to risk of litigation, no state/province or local agencies in the U.S. and Canada have established statutory requirements for minimum skid resistance, but most provide some guidelines [34]. The guidelines for skid number range from 30 to 40 for major highways (interstate and other highways with design speeds exceeding 65 km/h or 40 mph), while lower skid numbers are acceptable for low-speed and low-volume (average daily traffic less than 3,000) roads [93].

Surface texture can and should be part of the project level design process and many new technologies have been developed since [3]. For PCC pavements, macro-texture is provided by transverse grooves (with quality aggregate), transverse or skewed tining, diamond grinding (transverse or longitudinal), or longitudinal tining. Micro-texture in asphalt pavement can be produced through surface treatments (chip seals), open-textured friction course mixes, gap-graded mixes (SMA, NovaChip), and/or grooving.

10.4   Methods of Measuring and Reporting Skid Resistance

Prior to 1994, according to ASTME274, skid resistance was generally measured using a locked wheel skid trailer. This is still a widely used method in the U.S. and Canada.

In 1992, 16 countries participated in the international friction experiment under the auspices of PIARC. The purpose of the experiment was to compare and harmonize test results from various testing devices in use. The outcome of the experiment was the development of the International Friction Index (IFI), which, similar to the International Roughness Index (IRI), converts results from differing devices to a common scale. It directly linked surface friction to the macro-texture of the pavement.

The IFI requires the measurement of friction and macro-texture data and consists of two components: the Friction Number (F60) and the Speed Number (Sp). It is reported as IFI (F60' Sp) and is calculated as [87]:

$$F(S) = F(60) * e^{(60-S/Sp)}$$

where:
   $F(S)$ = adjusted value of friction for a slip speed of S
   $F(60)$ = measured friction value at a slip speed of 60 km/hr
   Sp = speed number, km/hr
   S = measured speed, km/hr
   MPD = mean profile depth (macro-texture), mm

The relationship between friction number, $F(60)$, Sp, and slip speed is shown graphically in Figure 10.4:

Historically macro-texture data has been measured using a volumetric technique. This basic method consists of spreading a known volume of material (sand, glass beads, or grease) on the pavement and measuring the

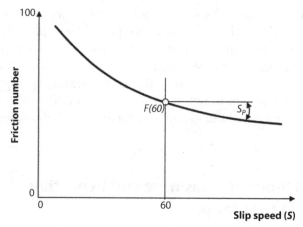

Figure 10.4 The IFI model. After [87]

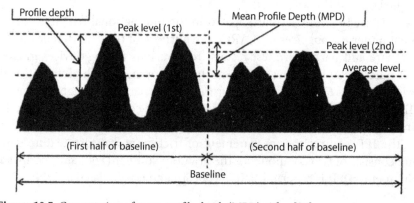

Figure 10.5 Computation of mean profile depth (MPD). After [87]

area covered. Dividing the volume by the area provides the Macro-Texture Depth (MTD).

The development of IFI has led to the potential for high-speed network level friction measurements as MTD and Mean Profile Depth (MPD), as illustrated in Figure 10.5 [87]. Friction and slip speed can thus now be measured using devices that do not require stop-start conditions.

10.4.1　Skid Measuring Equipment and Testing Protocols

Two protocols were prominent in 1994 for measuring skid resistance and surface texture: ASTM E274 for a calibrated locked-wheel skid trailer and

ASTM E303 and D3319 for the British Pendulum . The former measures the skid resistance of the pavement and the latter measures the aggregate polishing resistance.

With the development of the IFI, ASTM E274 was updated in 1997 (ASTM:E274-97) and several new methodologies for high speed data collection have been developed, as presented in Table 10.2. This table, excerpted from NCHRP 2009, is a summary of a larger one in the full report. A comprehensive comparison of all of the measurement methods, devices, and tools—including additional information necessary for planning friction testing programs, and the advantages and disadvantages of each approach—is found in the complete report [88].

10.5 Change of Skid Resistance with Time, Traffic, and Climate (Weather/Season)

The evaluation of skid resistance should recognize that changes will occur with time due to traffic, changes in the pavement surface, and seasonal changes in weather. Section 11.5, page 158, of [3] identifies these changes and the associated causes, all of which are still relevant today.

10.6 Including Friction Management in a Pavement Management System

With the development of high-speed friction measurement tools, agencies can now include friction measurement as part of network level pavement management. However, there are agencies still reluctant to fully embrace network or project level friction monitoring because of the potential liability that could arise if a section was known to have poor skid resistance and a fatality occurred between the time that the data was collected and a friction mitigation treatment was applied. In addition, the rapid change in skid resistance that occurs as a result of changing weather conditions and seasons may make network level friction management still difficult to justify for pavement managers.

Friction Management, like pavement management, is a systematic approach to measuring/monitoring friction and crash rates, identifying surfaces in need of remediation, and planning/budgeting for treatments and reconstruction. In [88] six key components of friction management are identified:

Table 10.2 Methodologies for friction testing. After [88]

Test Method	Associated Standard	Description	Equipment
Locked Wheel	ASTM E 274	Device is mounted on a trailer towed at a speed of 64 km/hr. Water 0.5mm thick is sprayed onto the road directly in front of the tire and a braking force is applied so that the tire locks up. Resistive drag force is measured and averaged for 1-3 sec. Measurements can be repeated and vehicle does not have to stop moving.	Tow vehicle and locked wheel trailer equipped with ASTM E 501 Ribbed tire or ASTM E 524 Smooth tire.
Side Force	ASTM E 670	Devices measure the pavement side friction or cornering force perpendicular to the direction of travel of one or two skewed tires. Water is placed on the pavement surface (1.2 L/min) and one or two skewed, free rotating wheels are pulled over the surface (typically at 64 km/hr). Side force, tire load, distance, and vehicle speed are recorded. Data is typically collected every 25 to 125 mm and averaged over 1 meter intervals.	The British Mu-Meter measures the side force developed by two yawed (7.5 degrees) wheels. Tires can be smooth or ribbed. The British Sideway Force Coefficient Routine Investigation Machine (SCRIM), has a wheel yaw angle of 20 degrees.
Fixed-Slip	various	Measure the rotational resistance of smooth tires slipping at a constant slip speed (12 to 20 percent). Water (0.5 mm thick) is applied in front of a retracting tire mounted on a trailer or vehicle typically traveling 64 km/hr. Test tire rotation is inhibited to a percentage of the vehicle speed by a chain or belt mechanism or a hydraulic braking system. Wheel loads and frictional forces are measured by force transducers or tension and torque measuring devices. Data are typically collected every 25 to 125 mm and averaged over 1 m intervals.	Roadway and runway friction testers (RFTs). – Airport Surface Friction Tester (ASFT). – Saab Friction Tester (SFT). – U.K. Griptester, – Finland BV-11. – Road Analyzer and Recorder (ROAR). – ASTM E 1551 specifies the test tire suitable for use in fixed-slip devices.

Variable Slip	ASTM E 1859	Variable-slip devices measure friction as a function of slip (0 to 100 percent) between the wheel and the highway surface. Water 0.5 mm thick is applied to the pavement surface and the wheel is allowed to rotate freely. Gradually the test wheel speed is reduced and the vehicle speed, travel distance, tire rotational speed, wheel load, and frictional force are collected at 2.5-mm intervals or less. Raw data are recorded for later filtering, smoothing, and reporting.	– French IMAG – Norwegian Norsemeter RUNAR – Road Analyzer and Recorder (ROAR).
Stopping Distance Measurement	ASTM E 445	Pavement is sprayed with water until saturated. Vehicle is driven at constant speed (64km.hr) and wheels locked. The distance the vehicle travels while reaching full stop is measured.	Can be a passenger car or light truck (at least 3200lb) with heavy duty suspension. The braking system should be capable of fully locking up. Tires should be ASTM E 501 ribbed design
Deceleration Rate Measurement	ASTM E 2101	Typically done during winter conditions when the surface is contaminated. While traveling at standard speed (32 – 48 km/hr) brakes are applied to lock the wheels, until deceleration rates can be measured. The deceleration rate is recorded for friction computation.	Mechanical or electronic equipment is installed on any vehicle to measure and record deceleration rate during stopping.
Portable testers	ASTM E 303 ASTM E 1911	Portable testers can be used to measure the frictional properties of pavement surfaces. These testers use pendulum or slider theory to measure friction in a laboratory or in the field.	

(Continued)

Table 10.2 Methodologies for friction testing. After [88] (*Continued*)

Test Method	Associated Standard	Description	Equipment
		The British Pendulum Tester (BPT) produces a low-speed sliding contact between a standard rubber slider and the pavement surface. The elevation to which the arm swings after contact provides an indicator of the frictional properties. Data from five readings are typically collected and recorded by hand. The Dynamic Friction Tester measures the torque necessary to rotate three small, spring-loaded, rubber pads in a circular path over the pavement surface at speeds from 5 to 89 km/hr. Water is applied at 3.6 L/min during testing. Rotational speed, rotational torque, and downward load are measured and recorded electronically. Results are typically recorded at 20, 40, 60, and 80 km/hr, and the speed/friction relationship can be plotted. The Tester fits in the trunk of a car and is accompanied by a water tank and portable computer.	
Electro-optic laser method	ASTM E 1845 ISO 13473-1, 13743-2, 13473-3	Non-contact high-speed lasers are used to collect pavement surface elevations at intervals of 0.25 mm or less. This type of system, therefore, is capable of measuring pavement surface macro-texture (0.5 to 50 mm) profiles and indices. Global Positioning Systems (GPS) are often added to this system to assist in locating the test site. Data collecting and processing software filters and computes the texture profiles and other texture indices.	High-speed laser texture measuring equipment (such as the FHWA ROSAN system) uses a combination of a horizontal distance measuring device and a very high speed (64 kHz or higher) laser triangulation sensor. Vertical resolution is usually 0.002 in (0.5 mm) or better. The laser equipment is mounted on a high-speed vehicle, and data is collected and stored in a portable computer.

Method	ASTM	Description	Equipment
Sand Patch Method (SPM)	ASTM E 965 ISO 10844	High-speed laser texture measuring equipment (such as the FHWA ROSAN system) uses a combination of a horizontal distance measuring device and a very high speed (64 kHz or higher) laser triangulation sensor. Vertical resolution is usually 0.5 mm or better. The laser equipment is mounted on a high-speed vehicle, and data is collected and stored in a portable computer.	Equipment includes: Wind screen, (25,000 mm3) container, scale, brush, and disk (60- to 65-mm] diameter). ASTM D 1155 glass beads.
Outflow Meter (OFM)	ASTM E 2380	Volumetric test measures the water drainage rate through surface texture and interior voids. It indicates the hydroplaning potential of a surface by relating to the escape time of water beneath a moving tire. Correlations with other texture methods have also been developed.	Equipment is a cylinder with a rubber ring on the bottom and an open top. Sensors measure the time required for a known volume of water to pass under the seal or into the pavement.
Circular Texture Meter (CTM)	ASTM # 2157	Non-contact laser device measures the surface texture in a 286-mm diameter circular profile of the pavement surface at intervals of 0.868 mm, matching the measurement path of the DFT. It rotates at 6 m/min and provides profile traces and mean profile depth (MPD) for the pavement surface.	Equipment includes a water supply, portable computer, and the texture meter device.
Stopping Distance Measurement)	ASTM E 445	Pavement is sprayed with water until saturated. Vehicle is driven at constant speed (64km.hr) and wheels locked. The distance the vehicle travels while reaching full stop is measured.	Can be a passenger car or light truck (at least 3200lb) with heavy duty suspension. The braking system should be capable of fully locking up. Tires should be ASTM E 501 ribbed design

(Continued)

Table 10.2 Methodologies for friction testing. After [88] *(Continued)*

Test Method	Associated Standard	Description	Equipment
Deceleration Rate Measurement)	ASTM E 2101	Typically done during winter conditions when the surface is contaminated. While traveling at standard speed (32 – 48 km/hr) brakes are applied to lock the wheels, until deceleration rates can be measured. The deceleration rate is recorded for friction computation.	Mechanical or electronic equipment is installed on any vehicle to measure and record deceleration rate during stopping.
Portable testers)	ASTM E 303 ASTM E 1911	Portable testers can be used to measure the frictional properties of pavement surfaces. These testers use pendulum or slider theory to measure friction in a laboratory or in the field. The British Pendulum Tester (BPT) produces a low-speed sliding contact between a standard rubber slider and the pavement surface. The elevation to which the arm swings after contact provides an indicator of the frictional properties. Data from five readings are typically collected and recorded by hand. The Dynamic FrictionTester measures the torque necessary to rotate three small, spring-loaded, rubber pads in a circular path over the pavement surface at speeds from 5 to 89 km/hr. Water is applied at 3.6 L/min during testing. Rotational speed, rotational torque, and downward load are measured and recorded electronically. Results are typically recorded at 20, 40, 60, and 80 km/hr, and the speed, friction relationship can be plotted. The Tester fits in the trunk of a car and is accompanied by a water tank and portable computer.	

1. Network definition
2. Network-level data collection
3. Network-level data analysis
4. Adequate monitoring of friction and crashes
5. Detailed site investigation
6. Selection and prioritization of short and long term restoration treatments

In addition, two threshold levels of friction management (similar to network and project level for PMS) were identified:

Investigatory–calls for detailed site investigation to determine need for remedial action. Action can include: erecting warning signs, identifying sections for more frequent testing, further analysis of friction and/or crash data, and application of short-term mitigation treatments (such as application of a chip seal).

Intervention–where because of extreme crash rates and/or low friction, immediate remedial action is required. Actions include: immediate restoration treatment, erection of warning signs, and a program of maintenance, rehabilitation or, construction.

A methodology to defining these levels is illustrated in Figure 10.6 [91] and may be site, region, or agency specific.

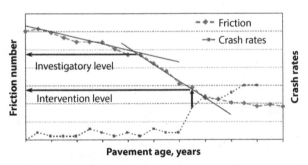

Figure 10.6 Establishing threshold levels for investigatory and intervention level friction management. After [91]

11

Combined Measures
of Pavement Quality

11.1 Concept of Combined Measures

Combined measures include the aggregation of individual measures. The
key issue is how to combine these measures and still indicate overall quality.

A combined measure or index of overall quality serves as a communication tool to senior administrators, elected officials, and to the public.
Individual measures, at the disaggregated level, are part of the technical and engineering decisions needed at the project level of pavement
management.

In general, the major individual measures that can be considered include
the following:

- Structural, usually in terms of deflection, or as a structural
  adequacy index
- Surface Condition in its many forms, or as a surface distress
  index
  a) rutting (permanent deformation)
  b) cracking (structural and environmental)
  c) roughness (present serviceability rating, present service-
     ability index, or international roughness index)

d) aesthetics

e) safety (usually measured by friction)

A measure of safety should not be part of a combined index unless there is provision to flag a safety issue that might be hidden or masked. Each of the foregoing factors or measures is important in determining pavement quality, but they are not equal and do not impact users or decisions equally or even at the same time. Some of these are "leading" indicators that point to future performance. Others are measures of current quality which define how well the pavement is performing right now.

Deflection is the most practical field measure of structural strength and is a "leading" indicator of future pavement distress and then performance. A pavement may appear to be in good condition at the present time, but if deflections are high in relation to the road class and traffic, it is likely to deteriorate rapidly in the future. In effect there will be a lag between the excessive deflection and the early onset of distress, roughness, or loss of serviceability. Depending on traffic and environment, high or increasing deflections may also point to the need to strengthen or upgrade the pavement to minimize future distress or early failure.

Surface distress on the other hand is a "lagging" indicator. If a pavement is rutted or cracked, it likely needs corrective maintenance to reduce safety hazards and to reduce the rate of rutting and cracking progression.

11.2   Examples of Combined Indexes

Roughness is the key measure that determines the present quality of the pavement. That is, how well it is serving the riding public today.

The most widely used indices that incorporate roughness include the Present Serviceability Index (PSI) [15], International Roughness Index (IRI) [94], and Riding Comfort Index (RCI) [95].

Among the various indexes that combine measures of surface condition or distress are the Pavement Condition Index (PCI), developed as a part of PAVER [96], and Surface Distress Index (SDI), developed in Canada [34].

Skid number, or friction on the other hand, is an independent indicator of how safely the pavement is serving the users (see Chapter 10). An unacceptable value is a direct indicator of the need for immediate maintenance.

Aesthetics, or how a pavement looks to the user, affects his/her perception of quality. A badly cracked pavement, for example, may be structurally sound, but the road user might perceive that it is worse than in reality. However, there has been little work on establishing an aesthetics index.

11.3   Developing Combined Indexes

Combined indexes can range from involving a group of measures, such as surface distresses with the result being a Pavement Condition Index (PCI) [96], to that of combining two or more indexes into an overall composite index. An early initiative, which has found widespread use, is the Pavement Quality Index (PQI) [97]. It incorporates a Structural Adequacy Index (SAI), a Surface Distress Index (SDI), and a Riding Comfort Index (RCI).

The key to building a composite or combined index is a structured technique for capturing expert opinion from a panel representing agency experience and management. In essence it is a calibration with the result being statistically robust and devoid of systematic errors [3]. As well, while the methodology itself can be transferable, it has to be specific to the agency.

While a composite or combined index can be extremely useful at the strategic or network levels, at the project level particular information can be masked in the aggregation. For example, a severe level of an individual distress requiring early corrective maintenance could be overlooked.

Another and similar caution is that of combining a leading indicator, such as deflection, with a lagging indicator, such as distress, without incorporating "trigger" values or limits into the process.

The PQI formulation described by [97] has found use in a number of states, provinces, and local agencies. One example that was developed in Minnesota is described in the next section.

11.3.1   Example Combined Index from Minnesota

This example, summarized from the work by [98], can be found in the MnDOT web site (www.MnDOT.org) with "An Overview of MnDOT's Pavement Condition Rating Procedure and Indices," May 9, 2006.

Minnesota uses three indices to report and quantify pavement condition. One represents roughness, one represents distress, and one the overall condition of the pavement, as shown in Table 11.1.

Table 11.1 MnDOT pavement condition indices

Index Name	Pavement Attribute Measured by Index	Rating Scale
Ride Quality Index (RQI)	Pavement Roughness	0.0–5.0
Surface Rating (SR)	Pavement Distress	0.0–4.0
Pavement Quality Index (PQI)	Overall Pavement Quality	0.0–4.5

The PQI is calculated from the RQI and SR as follows:

$$PQI = \sqrt{(RQI)(SR)}$$

Roughness or ride quality is quantified by the "serviceability-performance" concept developed at the AASHO Road Test. The serviceability of a pavement is expressed in terms of the Present Serviceability Rating (PSR). MnDOT refers to the present serviceability rating as the Ride Quality Index (RQI).

The first step in determining the RQI is to calculate the International Roughness Index (IRI) from the pavement profile.

While many states use the IRI as their sole measure of roughness, Minnesota converts it to RQI.

The correlation is done using a rating panel.

RQI categories and ranges are summarized in Table 11.2. Note that this is consistent with the original formulation of Present Serviceability Index (PSI) on a scale of 0 to 5 [15].

Using regression analysis, the panel's RQI is correlated to the measured IRI. Separate curves are established for bituminous and concrete pavements, as shown in Figure 11.1. The correlation is valid as long as the public's perception of smooth and rough roads does not change appreciably.

MnDOT uses the Surface Rating (SR) to quantify pavement distress. Their distress types and severities for bituminous surfaced, joined concrete, and continuously reinforced concrete pavements are captured by digital images and analyzed by operators at work stations. A 10% sample of each mile and station is taken. Weighting factors are used to calculate a Total Weighted Distress (TWD).

The TWD is correlated to the SR and details can be found in MnDOT's "Distress Identification Manual" (www.MnDOT.org). A graphical representation of the correlation is shown in Figure 11.2.

Table 11.2 RQI categories and ranges in MnDOT

Numerical Rating	Verbal Rating
4.1–5.0	Very Good
3.1–4.0	Good
2.1–3.0	Fair
1.1–2.0	Poor
0.0–1.0	Very Poor

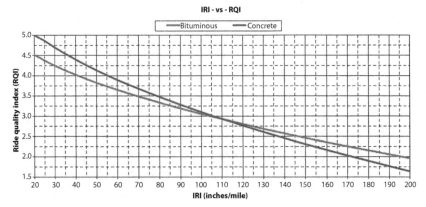

Figure 11.1 Conversion of IRI to RQI in MnDOT.

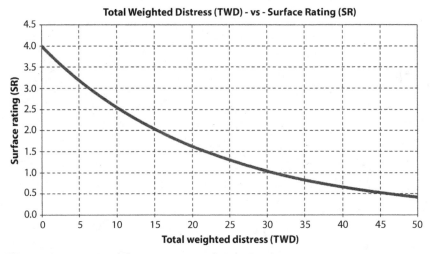

Figure 11.2 MnDOT chart or converting TWD to SR.

12

Data Base Management

12.1 Introduction

Access to and effective use of quality data is essential to pavement management. This means that the collection, processing, storage, and overall management of the data base has to be efficient and cost-effective [2].

The requirements for good data have not changed over the past decades, but what has changed are the technological, economic, and integrated asset management factors that characterize the present state of data base management. In the following sections these factors are discussed, as well as the key components or features of data base management systems.

12.2 Factors that Characterize the Present State of Data Base Management

Big data, metadata, smart networks, and "the cloud": none of these were on the horizon when pavement management systems were first implemented in the 1970s or even into the 1990s. Since then, improvements in information and communication technology (ICT) have proceeded at a dizzying speed. Forgetting for a moment the technological advances that have enabled collection, analysis, and reporting of pavement data with greater ease and sophistication, three factors have driven the need for more and more high quality data for pavement and infrastructure/asset management:

- Privatization of road networks and/or outsourcing of pavement and infrastructure management functions (design, maintenance, etc.),
- Widespread use of spatial data and the ability to visualize data using Geographic Information Systems (GIS), and
- The movement towards integrated asset management and a requirement for data to "do more."

Privatization and/or Outsourcing: The first agencies to privatize their road networks through contracts, concessions, public-private-partnerships, or other structures changed the way of doing business for road authorities and agencies around the globe. Overnight, agencies that privatized some or all of their management functions changed from being "deliverers of road networks," where all decisions regarding design, construction, maintenance and rehabilitation were made by an agency staff, to "keepers of the standards," where a smaller core staff managed standards for outside consultants and contractors to use. Depending upon the form of the contractual arrangement, multi-year funding decisions, performance bonuses/penalties, and/or maintenance and operation funding levels are determined by continuous monitoring of the road network by internal forces, external contractors, or a combination of both. Regardless of how it is collected, or by whom, a lot of data is needed for this business model and a well-designed, robust data base management system provides the foundation. Having to share data with multiple users means that agencies now have either to provide data in multiple formats that can be used by a variety of operating systems and programs or to make all the data open access via web portals. One example of open data access is Alberta Transportation that began publishing pavement condition data (IRI and rut depth at 50m intervals) on the Internet in 2003 (www.transportation.alberta.ca).

Spatial Data and Visualization: During the 1990s many agencies were building Geographic Information Systems (GIS) and GIS became the de facto data base integration mechanism for data from multiple asset categories. Linear networks such as roads, water distribution, wastewater and storm sewers, and fibre optic cables, as well as point objects such as bridges, signs, and light standards could all be overlaid on top of the base location file. The result was that linear referencing and analysis tools became more standardized and integrated. With the advent of Google Maps, visualization of data stepped into the mainstream and the final step for some agencies has been to put all of their infrastructure data onto Google Map. The city of Nanaimo in Canada made such a move in 2008 when it published all data to Google Earth, and now the public and contractors can view and/or download road, water, sewer, and other maps in various formats.

Integrated Asset Management: Data has always been an important and valuable asset, but with the evolution to asset management, as driven by GASB 34 in the United States, data has become even more important. All agencies now see data as a valuable asset to be guarded and robustly supported by dedicated business functions. Data no longer just supports multi-year investment decisions in terms of maintenance and rehabilitation: it now provides business intelligence and reporting in the form of performance measurement.

12.3   Some Evolutionary Features of Data Base Management

Having begun their journey into asset management with a pavement management system, for most agencies the foundational data base was, and still is, the pavement management system data base. State agencies in the United States began with the FHWA continuous highway condition reporting system in the 1960s that morphed into the Highway Performance Monitoring System (HPMS) in the 1970s. The approach used in HPMS formed the backbone for many state pavement management systems. The HPMS was comprised of three integrated major components: a data base system, the analytical process, and a reporting system. All of the pavement management systems that developed throughout the 1990s, whether stand-alone or integrated with other silo management systems (bridge, safety, etc.), had the same major elements, but with changes in the way pavements and other assets are managed, these components have become disentangled. The business model of road agencies has evolved to support separate analysis tools and user-specific, interactive reporting modules.

Municipalities followed suit and, depending on their size, either developed in-house custom designed systems, as in the case of the Municipal Pavement Management Application which is in use in seventeen cites in Alberta, Canada; or purchased commercial-off-the-shelf (COTS) systems; and/or contracted full-service turnkey systems provided by specialist consultants. Whether it is for a pavement section, bridge, culvert, light standard, sign, or any other asset within the right-of-way, all objects are defined by three types of data: what is it and where is it (inventory data), what are its characteristics (attribute data), and what condition is it in (performance data); and all objects are located in space (location referencing).

Changed business models and rapid advances in technology have made data base management an extremely specialized field. This chapter does not, in any way, attempt to provide detailed information on data base design and management. Rather, it is intended to be a road map for pavement managers who are new to the field, contemplating transforming their pavement data base into a more comprehensive asset management data base, considering outsourcing parts or all of their pavement management functions, and/or considering bringing their existing pavement turnkey system in-house.

Many of the references available are a result of symposia, best practice surveys, scanning tours, and various reports and other publications. There is still active discussion in the pavement community through international conferences (TRB Annual Conference, ICMPA 2008, 2011, and 2015, as well as GIS industry sponsored conferences and workshops). The topic area has evolved from pavement/road management systems, to integration through Geographic Information Systems, to asset management. Key references that illustrate this are [1,91,99–103].

There can be a difference between data (values that are recorded in a data base) and information (data that is understood by some user), but increasingly, data mining techniques have allowed us to add knowledge or wisdom to this continuum. In pavement management, the best example of the data into information process is the conversion of raw laser data into the International Roughness Index (IRI). IRI can then be categorized into a good-fair-poor rating (see the example in Chapter 11, Table 11.2). Asset management can, for example, take that rating and combine it with AADT to report on the percentage of population traveling on good, fair, or poor condition roads. This new measure informs policy makers at the highest level (usually the legislature) and becomes part of the macro-economic fabric of a province or state.

12.4   Data Base Management Systems and Key Components

Knowledge is an intangible asset that has incredible value. Beyond having information to make more effective decisions, being able to share data effectively and understand its accuracy and uses will allow us to make our current silos of information transparent. Nimble and dynamic systems leading to quick and accurate information is needed for effective decision-making. Data must only be collected once and used many times. Data must be understood and readily shared [91].

The terminology of data bases and data base management systems can be confusing to people not working in the information technology field. There are many excellent references, but for purposes of this chapter, some of the major terms are described in the following. An organization's data base management system can include many applications working at once and allow access through multiple computers. They may have many data bases within the overall architecture and data is connected through common key indices and/or location referencing.

A Data Base Management System (DBMS) has four components:

1. *Users* who use a data base to track things, use forms to enter, read, delete, and query data, and produce reports on the data base. Users are the business analysts and engineers who rely on the data base to make decisions (engineering, business, operations, etc.) using the data base applications.

2. *Database applications* are a set of one or more programs that serve as an intermediary between the user and the DBMS. Data base applications are programs that read and/or modify the data base using structured query language (SQL) statements to the DBMS and that present data to users in the format of forms or reports. Examples of data base application programs are Java, C#, HTML, and VBScript, amongst others. Data base applications have many functions, including creation and processing of forms and reports, processing of user queries, execution of application logic, and controlling of data base applications. The data base application is the core decision support system in a pavement management system that draws data from the data base and manipulates it to optimize multi-year maintenance and rehabilitation programs.

3. *DBMS* is the program used to create, process, and administer the data base. It receives requests encoded in SQL and translates these requests into actions on the data base. DBMS are complicated programs that are licensed software and are rarely custom written for corporations/agencies. Examples of DBMS are Oracle, DB2, and SQL Server. The functions

of a DBMS include: creation of the data base (tables and supporting structures), reading data base data, modifying data base data (insert, update, delete), maintaining data base structure, enforcing rules, controlling concurrency (allowing only one user at a time to update a data element, for example), providing security, and performing backup and recovery.

4. *Data base* is the collection of related tables containing columns, rows (referred to as "tuples"), and other structures. This is the base of the entire management system and is at its most simple a sequence of records with a fixed format stored in a single file. By breaking the record into multiple tables, data can be more easily accessed, retrieved, used, and updated. The two most commonly used data bases are relational data bases where multiples tables (comprised of rows and columns) are related through key indices (for example, a highway number, or a social security number) and object-oriented data bases where information is grouped into objects, so a road is the object and the sub-objects include information of the width, pavement type, layers, etc. All information on that road is attached to the object rather than being separately maintained in a table system. Object oriented data bases are niche systems that are sometimes layered on top of a relational data base.

Controlling all of this is the data base administrator who works with programmers and analysts to design and implement the data base, works with users and managers to establish policies, and implements security features and permissions for access, data creation, update, and deletion.

12.5   Advantages of Integrated Data Base Management Systems

There are many advantages of a data base management system, which include many individual but integrated data bases, as opposed to dedicated data bases attached to separate applications. These advantages include:

- program-data independence which allows for the applications to be changed and/or updated as new technologies develop,

- minimal data redundancy (input and update) which allows for separate data stewardship functions whose entire function is to receive QA and QC data before publication to the data base,
- improved data consistency and quality which is increasingly important as agencies rely on contractors who supply data and may change on an annual basis,
- improved data sharing between and within organizations, which is particularly useful in urban areas where utilities overlap and coordination of rehabilitation/maintenance works can be facilitated with shared data,
- increased productivity and application development/ implementation,
- enforcement of data standards,
- improved data accessibility, and
- reduced program maintenance.

12.6   Examples of Integrated Data Base Management

An example of an integrated highway data base is illustrated in Figure 12.1. The Transportation Information Management System (TIMS) of Alberta Transportation is owned and operated by data architects, analysts, and stewards who do not play a role in decision-making; rather, they receive, assure quality, and control data provided by data contractors prior to publishing it to the web. The data base management system is fully normalized, which means that every data point is separately located to the highway network so that, for example, one wheel path rut depth value is separate to the adjacent wheel path and a data steward can QA/QC it using computerized reliability checks. This form of data warehouse presents some performance issues when it comes to analysis; therefore, the data is de-normalized in advance of its use for the provinces pavement management application. What de-normalizing means is that prior to analysis, every rut depth value (left and right) are joined and located to the highway section. This makes more sense for development of network level multi-year programs, but in its raw, normalized form, the data can be used for identification of project level micro-surfacing projects where one wheel path is rehabilitated and the other not.

At its core TMIS has a data repository that generates reports through a communication application (shown at the top of Figure 12.1) and both feeds data to and receives data from many applications (PMA, RODA, NPEDA, GIS, BIS, etc.).

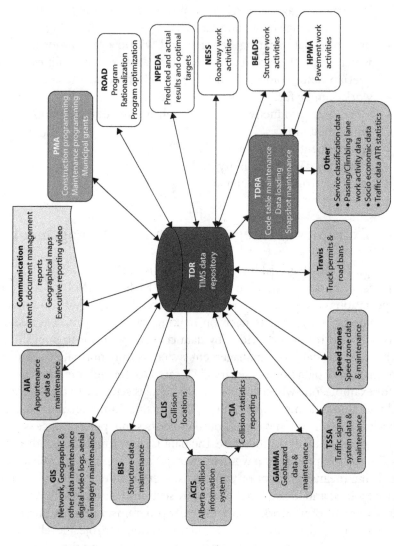

Figure 12.1 Alberta's Transportation Information Management System (TIMS) data base design. After [95]

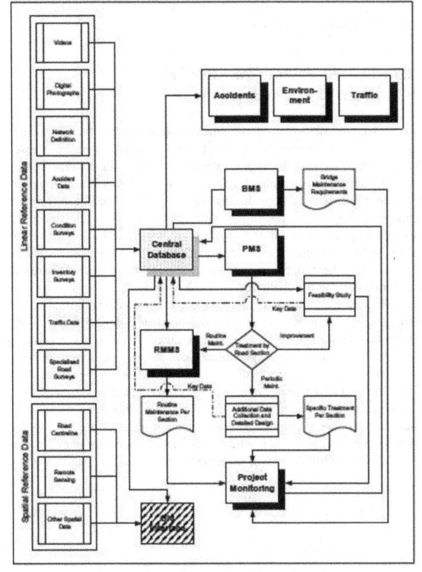

Figure 12.2 Example of a Road Management System (RMS) framework. After [101]

The World Bank [101] also provided a framework for an integrated road management system as illustrated in Figure 12.2. In this framework, the central data base has data coming to it from linear and spatial referencing sources and then has several applications drawing data for analysis, namely the PMS, BMS, RMMS, etc.

Both of these models illustrate the potential complexity of DBMS, and there is no "one size fits all" model.

12.7   Success Factors for Effective Data Base Management

As discussed in [3], successful implementation of pavement/asset management requires a triumvirate of people, technology, and business processes. Figure 12.3 illustrates this clearly: without any one of these elements supported by appropriate funding levels, no management system will be successfully implemented. With disentanglement of the data base from the analytical process and reporting systems, decreasing data collection intervals, and huge quantities of data, it is doubly important for the data warehouse/stewardship and data mining functions to be properly supported.

Confirming the importance of the triumvirate, [91] presented the results of a symposium on the challenges of using existing data base management systems for performance measurement within the context of broader asset management. Agencies were quite varied in their approach to data base management, but many saw it as a separate function with separate business plans that aligned with the agency mission, strategic and tactical plans, dedicated data stewards, performance metrics, and technology investment plans. The data base management system is clearly seen as a strategic asset that requires management of its own. Like materials, energy, and human resources, data are an important asset for planning, building, and operating transportation systems, public and private. Data cost money and can provide commensurate returns on investment. System managers need to

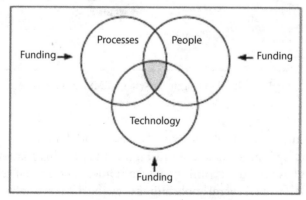

Figure 12.3 The triumvirate for successful road management. After [101]

plan for and allocate resources to collecting and maintaining data bases sufficient in coverage, quantity, and quality to support transportation decision making [102].

Paterson and Scullion [99] defined three success factors for data base management that are still relevant regardless of the type of data being collected and/or how it is being stored. Specifically, data should be relevant, reliable, and affordable.

- *Relevance* means that all data being collected/stored should have a use and that collecting/storing data for data sake should be avoided. This is particularly an issue when a system is evolving from a "silo," such as a pavement management system, to an integrated asset management system. Migrating legacy data can be a painful, costly, and time-consuming experience. Nonetheless, it is important to review data needs periodically to prevent calcification of data collection methods and protocols. For example, in 1997/98 a review of the original HPMS system eliminated 15 data elements and substantially altered 21 others. At that time the HPMS data base was valued at an estimate $15–20 million USD.
- *Reliability* is a difficult balance between accuracy, precision, coverage, completeness, and currency. Reliability goes hand in hand with responsibility and it must be clear who is responsible for collecting, processing, and publishing data. With rapidly changing data collection methods (as discussed in earlier chapters), there is a risk that credibility can be lost if transformed data (such as IRI) is found to have higher accuracy as a result of technological improvement. Similarly, as data collection is outsourced QA and QC processes must be in place and robust to ensure that published data is as accurate as possible [2]. Data coverage is one of the elements that has seen the greatest change in the past couple of decades as the technology of pavement roughness and strength measurement now allows for continuous measurement using lasers and Doppler radar. With the proliferation of smart phones and tablets, "citizen raters" can now collect data on behalf of agencies and reliability may become a challenge for pavement managers. The City of Calgary 311 system encourages the public to send photos and reports of potholes, sidewalk tripping hazards, and other maintenance needs directly

to their GIS data base. Similarly, Parks Canada is developing a prototype system whereby hikers can snap a QR code with their smartphone and send a photos of a trail bridge or other back-country facility that needs maintenance/rehabilitation.

- *Affordability* can be the biggest constraint to data base design not only in terms of the data base architecture, information, and communications technology supporting it, but also in the data that will populate it. Technology is changing the picture as advances in data collection technologies—such as real-time tracking of vehicles and shipments and monitoring infrastructure components, Internet-based survey methods, remote surveillance, video imaging and interpretation, and cellular phone-based data collection—are making it easier to collect more and more accurate data about transportation and travel. These innovations can improve decision support, but care is needed to avoid swamping the decision process with data. Concerns about personal and business privacy will also need to be addressed [104].

13

Communicating the Present Status of Pavement Networks

13.1 Introduction

Pavement Managers need to be able to respond to the following basic questions:

1. What assets do we have and where are they located?
2. What condition are they in now and estimated to be in the future?
3. What is their current worth?
4. What do we need to do and how much will it cost to maintain the current condition?
5. What happens if we do not receive the funds necessary to maintain the current condition?

Once agency managers determine the condition of their pavement assets, they must communicate that information to their stakeholders in an appropriate way. Since the early days of pavement management, agencies have developed robust reporting mechanisms, including "State of the Pavement Reports" and interactive maps using Geographic Information

Systems (GIS). The level of detail of these reports varies for the audience, ranging from sectional IRI data for the technical pavement engineers, to green-yellow-red color coded maps for the non-technical public, to balanced scorecard report cards for all infrastructure elements for legislators/policy makers.

The move to integrated and comprehensive infrastructure asset management in many jurisdictions (as a result of GASB 34 in the U.S. and the Tangible Capital Asset Reporting requirement in Canada, for example) has expanded the scope and importance of reporting the condition of pavement networks. In particular, the emphasis has shifted away from pure condition reporting to performance measurement.

The following resources can be explored for more information on how to develop performance measures:

- Transportation Research Board Annual Conferences, 1998–2002, including specialty conferences on Managing Pavements, Asset Management, Low Volume Roads and Performance Indicators
- FHWA Office of Asset Management source documents
- Transportation Research Board on-line state sponsored websites
- World Bank documents
- Transportation Association of Canada documents

13.2 Performance Measures

Performance measures are used by agencies to 1) define policy objectives at an early stage of policy or system planning, 2) provide the basis for annual performance reporting on system condition and performance as part of communications, 3) screen projects or set priorities, and 4) allocate resources [105]. Performance measures should be defined in response to the goals and objectives, which are directly aligned with the broad goals and mission of the agency. To be effective, performance measures should be:

- Based upon technically sound, repeatable, robust data, which is supported by the agency business processes.
- Understandable to all levels of the agency (technical, administrative, and executive) and capable of being "rolled up" for non-technical reporting to the public at large.

- Reflective of the user or stakeholder groups: in the case of transportation agencies, the public and commercial users, the service providers (the department of highways and transportation), and policy makers (executive branch and/or cabinet).
- Broad enough to sum up the net effect at the system level of many smaller, discrete actions, but specific enough, at the technical level, to register a response to decisions (that is, a change in the decision causes a response in the measure and "moves the needle") [105].

Stakeholder involvement is an important consideration when developing outcome based performance measures, requiring an understanding of the needs, expectation, and levels of satisfaction of the users or customers. According to the Transportation Association of Canada (TAC), the group of stakeholders that should be involved in or considered when developing and establishing performance measures are the following [106]:

- Highway agencies as service providers, consultants acting on their behalf, and contractors who have taken over network roads in long term performance based contracts. Also included could be the supplier of goods and materials and financing agencies, such as electronic toll roads).
- Private and commercial road users, such as drivers of cars, trucks, buses, etc.; motorcyclists and bicyclists; and pedestrians.
- Policy makers and regulators, particularly regarding issues such as fuel taxes and tools in the case of policy makers, and compliance with road laws, safety, and vehicle weights and dimensions in the case of regulators.
- The public at large to whom the public agency is accountable for highway performance.

13.3   Performance Measurement and Strategic Level Pavement Management

Performance measures operate on three levels within an agency and must be interrelated and capable of being rolled up. Two levels at which pavement management must operate, network and project levels, were defined in the early stages of pavement management [107]. Asset management adds a third level, that of strategic management which aligns the network and project level goals and objectives to the corporate policy of the agency. It therefore follows that performance measures, as part of an asset management system, should operate at three levels: strategic, network, and project. Strategic level performance measures are defined within the Business Plan of the agency and address the highest goals and objectives. Strategic level measures must span asset categories (that is, roads, bridges, appurtenances, lighting, etc.). An example of a strategic level performance measure within a highway agency is crash rates. By setting a target for reduction in crashes, the agency must provide a safe highway system in terms of capacity, geometrics, and road surface characteristics, and as such, it presents an umbrella indicator. Similarly, a strategic objective such as "provide economic growth" requires action from many departments through various government-wide initiatives. Based on the example of crash rates, regulations such as graduated licenses and minimum vehicle safety standards can be introduced to also contribute to lower crash rates through better drivers and fewer mechanical hazards.

Strategic level performance can also be reported or communicated at the macro-level in terms of performance measures that are understandable to the general public, as shown in Table 13.1, a summary from [108-110]. Another, more specific measurement of service quality provided to road users, summarized from the same references, is provided in Table 13.2. These tables are also contained in [111].

Performance measures should be defined in response to the goals and objectives that are directly aligned with the broad goals and mission of the agency as illustrated in Figure 13.1.

There are two different approaches to translating long-term goals and objectives into specific performance goals for use in planning and programming as illustrated in Figure 13.2. In the prospective approach the goals are established and plans put in place to achieve them, while in the retrospective approach, the plans are defined and the goals are derived from the existing plans.

13.4   Performance Measure Categories

Performance measures generally fall into three categories: inputs, which look at the resources dedicated to a program (such as the dollars spent, materials, or staff time consumed); outputs, which look at the products produced (such as miles of pavements placed, miles of lanes added); and outcomes, which look at the impacts of the products on the goals of the

Table 13.1 General, macro-level performance measures for key road assets. After [108,109,112]

Feature or Aspect	Indicator	Units	Breakdown and Remarks
1. Network size or extent	a) length	center line-km and lane-km	By road class, jurisdiction, urban or rural
	b) paved/ unpaved	% and length	By road class, jurisdiction, urban or rural
	c) bridges, culverts	Number	By type or category
	d) tunnels	Number	By type or category
	e) links	Number	Ferries, road-rail, etc.
	f) right-of-way area	Ha	
2. Asset value	a) replacement	$	By indicators in 1
	b) book value or written down replacement cost	$	By indicators in 1
3. Road users	a) registered vehicles	Number	By cars, SUV's, light trucks, classes of heavy trucks, buses, motorcycles, etc.
	b) ownership	Vehicles / No. of owners	By cars, SUV's, light trucks, classes of heavy trucks, buses, motorcycles, etc.
	c) trip purposes	Trips, person-km, or vehicle-km	By work, recreational, commercial, etc. categories

(Continued)

Table 13.1 General, macro-level performance measures for key road assets. After [108,109,112] (*Continued*)

Feature or Aspect	Indicator	Units	Breakdown and Remarks
4. Demography and macro-economic aspects	a) Population	Number	
	b) total land area	Sq. km	By climate, topography, region, etc.
	c) urbanization	% of population	
	d) GNP or GDP	Total $	Also $/capita
5. Network density and availability	a) road density	Km/ 1,000 sq. km	
	b) road availability	Km/10^6 persons	
6. Utilization	a) travel	Veh-km/yr	By road & vehicle class, dollar value
	b) goods	Tonne-km/yr	
7. Safety	a) accidents	Total no. and rate	Rate in terms of no./ 10^6 veh-km
	b) fatalities	Number	Rate in terms of no./ 10^6 veh-km
	c) injuries	Number	Rate in terms of no./ 10^6 veh-km

Table 13.2 Measures of service quality provided to road users. After [108,109,112]

Feature or Aspect	Indicator	Units	Breakdown and Remarks
1. Comfort/ Convenience	a) ride quality	IRI, RCI, etc.	Clear definitions of units and methods are essential
	b) surface quality	rut depths, IFI, SN, shoulder types and widths	Clear definitions of units and methods are essential
2. Road corridor	a) geometrics	grades, curvature, lane, widths, cross slopes, sight distance	% radii or degrees for grades and curvature., m for lane widths and sight distance
	b) driver guidance	markings, signs, messages	Locations, comprehension or awareness, legibility
	c) hazards	barriers, obstacles, distractions	Locations and numbers
3. Safety Risk	a) fatality	fatalities/ 10^6 veh-km	
	b) injury	injuries/10^6 veh-km	
	c) accident	total accidents /106 veh/km	
4. Mobility and Speed	a) delays	veh - hrs	
	b) congestion	% veh/km	Classified by adequate, tolerable and unacceptable for % of veh/km
	c) ave. travel speed	km/h	By road class, urban and rural
	d) closures	no. of days	By road link and causes
	e) clearance and load restrictions	no. of violations of standards, no. of trucks detoured, detour user cost	Primarily affects trucks

(Continued)

Table 13.2 Measures of service quality provided to road users. After [108,109,112] (*Continued*)

Feature or Aspect	Indicator	Units	Breakdown and Remarks
5. User costs	a) vehicle operating costs	average $/veh-km	For existing conditions
	b) travel time costs	$/veh-km	
	c) accident costs	$/10^6 veh-km	
6. Time Reliability	a) standard deviation of travel time		Often based on sample trips and reported by corridor
7. Environment	a) emissions	kg/10^6 veh-km	By hydrocarbon and other compound type
	b) noise	dB variation with time	Site Specific
8. Operational effectiveness	a) incident response time	minutes	Ave. by incident
	b) claims	$	Due to potholes or other unrepaired problems
	c) injury response time	days	Time to reply to inquiries or complaints

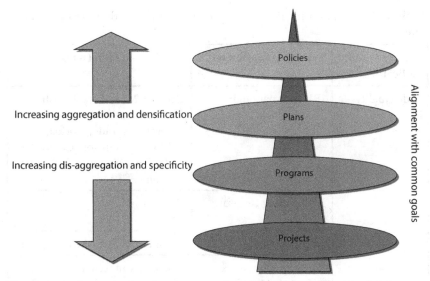

Figure 13.1 Alignment of performance measures with common goals. After [105]

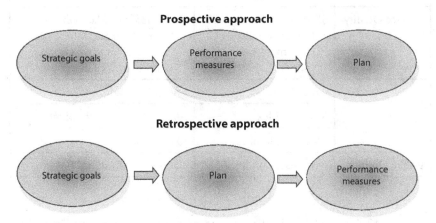

Figure 13.2 Translating goals into performance measures. After [105]

agency (such as discernible improvements in pavement ride, reduced travel time). Developing outcome based performance measures requires an understanding of the needs, expectations, and levels of satisfaction of the users or customers. A review of current state of practice performance measurements in North American found that most performance measure in use by North American transportation agencies fall into one of three major classes: Condition, Functional Adequacy, and Utilization. Table 13.3 provides a summary of performance measures for Condition.

Table 13.3 Condition based performance measures in North America DOTs. After [111]

Service Quality Indicator	Type of Measurement	North American DOTs
Ride Quality	IRI (International Roughness Index)	Ontario, Minnesota, British Columbia, Washington, Idaho, Iowa, Colorado, Florida, Montana, Oregon, California, Arizona, Michigan, Nebraska, Utah
	RCI (Riding Comfort Index)	Ontario
	RI (Ride Index)	Colorado, Idaho, Michigan, Utah
	PSI (Present Serviceability Index)	Arizona, Minnesota
	RN (Ride Number)	Florida
Surface Quality	SR (Surface Rating), DI (Distress Index), PCI (Pavement Condition Index)	California, Florida, Idaho, Iowa, Michigan, Minnesota, Nebraska, Ontario, Oregon, Washington, Utah
	Crack Index	Idaho, Colorado
	Rutting Index	Idaho, Colorado, Washington
	SN (Skid Number)	Idaho, Utah
Public Opinion	Pavement Condition	California, Montana, Ontario

Table 13.4 presents utilization based performance measures in use by North American transportation agencies.

How these performance measures are reported varies widely across North America. Florida reports Level of Service (LOS) as % of the network meeting a target LOS, while Idaho reports the % of the network exceeding the target LOS. Some agencies combine measures. For example, Florida reports the % of the daily vehicle miles travelled (VMT) at a target LOS. Alberta Transportation annually reports to the legislature the average International Roughness Index for the primary highway network and has studied the potential for incorporating vehicle miles traveled into the performance measure. The intent is to report the number of VMT on roads

Table 13.4 Utilization based performance measures in North American DOT's. After [111]

Service Quality Indicator	Type of Measurement	North American DOT's
Delay	LOS (Level of Service) – Volume/Capacity Ratio	Arizona, British Columbia, California, Colorado, Florida, Idaho, Iowa, Michigan, Minnesota, Oregon
	CI (Congestion Index)	Montana
	TRI (Travel Rate Index)	Colorado, Washington
	Real travel times	Washington
	Commercial Trucking travel times	British Columbia, Florida
	Congestion due to incidents	Washington
VMT (Vehicle Miles Travelled)	Annual VMT	Colorado, Florida, Michigan, Oregon, Utah
	Daily VMT	Montana, Oregon, Utah, Washington
Environmental	Emissions	California, Michigan, Oregon, Utah

that are in fair condition or better. The definition of "fair or better" as a specific technical value may be difficult to understand or interpret to a lay person, but most people understand the generalized concept of "fair or better." Regardless of which performance measure is used, it is important that the data be available and supported by the agency.

Table 13.5 presents functional adequacy based performance measures in use by North America departments of transportation and highways.

13.5 Example Report on the State of a Road Network in Terms of International Roughness Index

This example involves a real life, quite large network comprising two road classes (interurban or primary highways and rural or secondary highways),

Table 13.5 Functional adequacy based performance measures in North American DOTs. After [111]

Service Quality Indicator	Type of Measurement	North American DOTs
Design Geometrics	Considers geometrics of highway-including design characteristics, width, speed and obstructions	California, Florida, Idaho, Iowa, Michigan, Minnesota, Nebraska
Mobility	Load Restrictions	Idaho, Minnesota
	Accessibility	Ontario
	Level of Development	Arizona, California, Florida
Safety	Collisions – Fatalities	British Columbia, Colorado, Florida, Idaho, Iowa, Michigan, Minnesota, Montana, Nebraska, Ontario, Oregon, Utah
	Collisions – Injuries	Colorado, Florida, Idaho, Iowa, Michigan, Minnesota, Montana, Nebraska, Oregon, Utah
	Safe commercial vehicles and carriers	Ontario

1,293 pavement sections (spanning 3,240 center line km), 161 bridges, 356 culverts, and 45 major signs. A comprehensive data base was available, as well as rehabilitation and major maintenance treatment options, deterioration/ performance prediction models for the pavements, expected service lives for the other than pavement assets, unit costs, vehicle types and volumes, vehicle operating costs, etc. [112].

Figure 13.3 (a and b) shows a current "snapshot" of the state of the pavement network in terms of the performance measure/metric/indicator being International Roughness Index (IRI). The distribution is again an aggregation of detailed data into four classes of IRI. Stringent limits have been set for this network with IRI ≤ 1.0 as excellent, 1.5 ≥ IRI > 1.0 as good, 2.0 ≥ IRI > 1.5 as fair and IRI > 2.0 as poor. Thus the state of this infrastructure can be considered as follows:

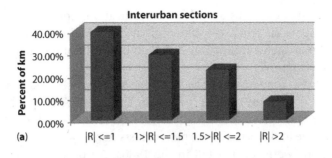

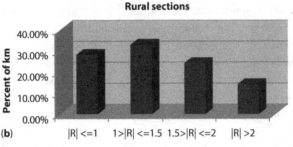

Figure 13.3 (a) Example state of interurban sections of a 3,240 km network. (b) Example state of rural sections of a 3,240 km network. After [112]

- Interurban: more than one third excellent, about one quarter good, one fifth fair, and less than 10% as poor.
- Rural: about one quarter excellent, one third good, one fifth fair, and a little more than 10% as poor.

13.6   Example Report on the State of a Road Network in Terms of Asset Value

This example involves asset value as a measure of assessment for state of the road infrastructure. Asset valuation is a complex subject, and the method(s) used can provide widely varying results. For instance, book value/historical cost commonly used in financial accounting (e.g., a past based method) versus written down replacement cost used by some agencies as a management accounting approach (e.g., a current based method) can illustrate these varying results. Similarly, a change in asset value over time is a more meaningful measure than only a current value viewed in isolation.

The example network of 113 sections is described in detail in [113], including cost data, performance models, etc. A base year of 1993 was used as the "current year" and predictions were made for 1999, as a "future year," for which actual data was available for verification. The purpose was not only to compare current asset values but also to predict future asset value as an asset management function and as an indicator of the changing state of the infrastructure.

A summary comparison of the "future year" predicted and actual/measured value of the network is provided in Table 13.5. The numbers illustrate a high book value/ historical cost (BV/HC) compared to the others, which was due to distorted construction costs during a boom period when these sections were constructed in the 1980s. An opposite situation can also occur, which suggests that book value/historical cost should be viewed with considerable caution in valuing pavement assets.

Written down replacement cost (WDRC) for current value (base year) is comparatively much lower because the construction costs had decreased by about one third or more from the 1980s' BV/HC bases.

The values in Table 13.6 and the much more comprehensive analysis in [113], which included statistical significance tests, suggest the following:

- Agencies who are carrying out asset valuation need to clearly recognize that considerable variation can exist between

Table 13.6 Example assessment of the state of a road network in terms of asset value (Current and Future). After [113]

Method	Base Year Current Value ($ million)	"Future Year" Predicted Value ($ million)	Actual (Measures Future Year Value	Difference (Predicted Value Measured)
BV/HC	155	155	155	0
WDRC	46	–	–	–
RC	81	113	105	8 (8%)
WDRC(SL)	–	72	67	5 (8%)
NSV$_a$	–	104	96	8 (9%)
WDRC(Eng)	–	64	53	11 (21%)
NSV$_b$	–	91	71	20 (28%)

Notes: BV/HC = Book Value/Historical Cost; WDRC = current written down replacement cost; WDRC(SL)=WDRC based on a financial straight line model; NSV$_a$ = net salvage value using a simple decision tree for rehabilitation; NSV$_b$ = NSV using a multi-point decision tree; WDRC(Eng.) is WDRC based on an engineering deterioration model; RC = replacement cost.

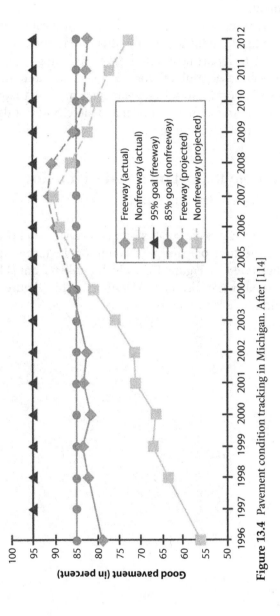

Figure 13.4 Pavement condition tracking in Michigan. After [114]

methods, particularly past based vs. current, BUT it is dangerous to generalize from one situation (e.g. the example Table 13.6) to another era or jurisdiction or infrastructure element.

- If asset value is used as a performance indicator for assessing state of the infrastructure, it is important that agencies are able to report how well they are retaining or improving asset value as a result of proper management and funding. It is also important to select a valuation method that is easily sustained and managed, understandable, and not data and/or analytically burdensome.

13.7   Example Report on a State Timeline of "Good" Pavement

The State of Michigan, as part of its system preservation efforts, carries out a time line projection of pavement condition in terms of the percent "good," as illustrated in Figure 13.4, which is taken from [114] where the terminology used is "Performance Measures." While Figure 13.4 does not contain minimum levels of performance, the 95% and 98% goals used are similar in concept to implementation targets.

References for Part Two

1. Flintsch, G. and K. K. McGhee, "Quality Management of Pavement Condition Data," NCHRP 401, Transportation Research Board of the National Academies, Washington, D.C., 2009.
2. Pierce, Linda M., G. McGovern and K. A. Zimmerman, "Practical Guide for Quality Management of Pavement Condition Data Collection," U.S. Department of Transportation, Federal Highway Administration, Feb., 2013.
3. Haas, R., W.R. Hudson, J.P. Zaniewski, *Modern Pavement Management*, Krieger Publishing, Florida, 1994.
4. Marsden, G, C.E. Kelly, C. Snell, "Selecting Indicators for Strategic Performance Management," Transportation Research Record 1956, 29-29, ISSN 0361-1981, 2006.
5. Bolstad, Paul, *Gis Fundamentals: A First Text on Geographic Information Systems*, 3rd Edition, Ingram Publishing, May 31, 2007.
6. Miller, H.J. and S.L. Shaw, "Geographic Information Systems for Transportation: Principles and Applications (Spatial Information Systems)," ISBN-10: 0195123948, ISBN-13: 978-0195123944, Oxford University Press, November 22, 2001.
7. Roads & Bridges Magazine, "Asset Management: Better Reporting Utah DOT's Data and Planning Tools Provide a Guide to the Future and Power of Network-Level Asset Data," Article by Robert Dingess, January 15, 2014.
8. ESRI web page: http://support.esri.com/en/knowledgebase/GISDictionary/, accessed 2014.
9. White, B and A. Rocco, "GIS and Pavement Management: A Concrete Relationship," Carson, California, ESRI web page: http://proceedings.esri.com/ library/userconf/proc98/proceed/to350/pap345/p345.htm, 2014.
10. Carey, W.N. and P.E. Irick, "The Pavement Serviceability-Performance Concept," HRB Bulletin 250, Highway Research Board, 1960.
11. Rasmussen, R.O., H.N. Torres, R.C. Sohaney, and S.M. Kramihas, "Real-Time Smoothness Measurements on Portland Cement Concrete Pavements During Construction," Report S2-R06E-RR-1, SHRP 2 Renewal Research, Transportation Research Board, 2013.
12. Van Geem, C. and B. Beaumesnil, "Comparison of Different Evenness Measurements, Application on the Case of a Newly Constructed Road Section," 7th Symposium on Pavement Surface Characteristics: SURF, Virginia Tech Transportation Institute, 2012.
13. Sayers, M.W., T.D. Gillespie, and W.D.O. Paterson, "Guidelines for Conducting and Calibrating Road Roughness Measurements," World Bank Technical Paper Number 46, Washington D.C., 1986.
14. Wisconsin Department of Transportation, "Pavement Roughness (IRI) REPORT: 2010," WisDOT, 2012.
15. Carey, W., and P. Irick, "The Pavement Serviceability – Performance Concept," Highway Research Board Special Report 61E, *AASHO Road Test*, pp 291-306, 1962.

16. FHWA flyer, Update of TPF- 5(063) "Improving the Quality of Pavement Profiler Measurement," 2003.

17. Wang, H, "Road Profiler Performance Evaluation and Accuracy Criteria Analysis," Thesis, Virginia Polytechnic Institute and State University, Blacksburg, Virginia, 2006.

18. Wang, H. and G.W. Flintsch, "Profiler Performance Evaluation and Accuracy Criteria Analysis," TRB 86th Annual Meeting Compendium of Papers CD-ROM, 2007.

19. Transtec, ProVal software web page, http://www.roadprofile.com/, accessed 2014.

20. Karamihas, S. M., "Critical Profiler Accuracy Requirements," University of Michigan Transportation Research Institute Report UMTRI-2005-24, 2005.

21. Karamihas, S.M., "Improving the Quality of Pavement Profiler Measurement," Benchmark Test Evaluation Report, Federal Highway Administration Project, University of Michigan Transportation Research Institute, 2011.

22. Face Dipstick, http://www.dipstick.com/store.php/dipstick/pg8749/road_profiler_info, accessed 2014.

23. APR Consultants, http://www.aprconsultants.com/The-Auto-Rod-and-Level.html, accessed 2014.

24. Surface Systems and Instruments, Inc., http://www.smoothroad.com/products/walking/, accessed 2014.

25. International Cybernetics, http://www.surpro.com/, accessed 2014.

26. Smith, H. and J.B. Ferris, "Calibration Design and Accuracy Verification of High Fidelity Terrain Measurement Systems," Journal of Testing and Evaluation, Vol. 38, No. 4, pp. 431-438, 2010.

27. Smith, H., and J.B. Ferris, "Techniques for Averting and Correcting Inertial Errors in High-Fidelity Terrain Topology Measurements," Journal of Terramechanics, Vol. 47, No. 4, DOI: 10.1016/j.jterra.2010.04.002, pp. 219-225, 2010.

28. Winkler, C.B., S.M. Karamihas, M.E. Gilbert, and M.R. Hagan, "Benchmark Profiler Field Manual, Version 2.3," University of Michigan Transportation Research Institute, 2013.

29. MnDOT, "Inertial Profiler Certification Program," Minnesota Department of Transportation, 2011.

30. Olson, M.J. and A. Chin, "Inertial and Inclinometer Based Profiler Repeatability and Accuracy Using the IRI Model," Rpt FHWA-OR-RD-13-03, Oregon State Univ, 2012.

31. Minnesota Department of Transportation, "Pavement Distress Identification Manual," Office of Materials and Road Research Pavement Management Unit, July 2011.

32. NCHRP, "Comparative Performance Measurement – Pavement Smoothness," AASHTO Standing Committee on Quality Report 2008.

33. Al Omari, B., and M.I. Darter, "Relationships between International Roughness Index and Present Serviceability Rating," Transportation Research Record 1435, 1994.

34. TAC, "Pavement Design and Management Guide," TAC, 1997.

35. Sayers, M.W. and S.M. Karamihas, "Estimation of Rideability by Analyzing Longitudinal Road Profile," Transportation Research Record 1536, 1996.

36. Edwards, L. and Q. Mason, "Evaluation of Nondestructive Methods for Determining Pavement Thickness," U.S. Army Corps of Engineers Report No. EDRC/GSL TR-11-41, Geotechnical and Structures Laboratory, Vicksburg, Mississippi, 2011.

37. Al-Qadi, I.L., S. Lahouar, K. Jiang, K.K. McGhee, and D. Mokarem, "Accuracy of Ground-Penetrating Radar for Estimating Rigid and Flexible Pavement Layer Thicknesses," Transportation Research Record: Journal of the Transportation Research Board, No. 1940, Transportation Research Board of the National Academies, Washington, D.C., 2005.

38. Cao, Y., S. Dai, J.F. Labuz, and J Pantelis, "Implementation of Ground Penetrating Radar," Minnesota DOT Report No. MN/RC 2007-34, St. Paul MN, 2007.

39. Cao, Y., B.B. Guzina, and J.F. Labuz, "Pavement Evaluation Using Ground Penetrating Radar," Minnesota DOT Report No. MN/RC 2008-10, St. Paul MN, 2008.

40. Holzschuher, C., H.S. Lee, and J. Greene, "Accuracy and Repeatability of Ground Penetrating Radar for Surface Layer Thickness Estimation of Florida Roadways," Florida DOT Report FL/DOT/SMO/07-505, Florida, 2007.

41. Infrasense, Inc., "Feasibility of Using Ground Penetrating Radar (GPR) for Pavements, Utilities, and Bridges," South Dakota DOT Report No. SD2005-05-F, Pierre SD, 2006.

42. Liu, W. and T. Sculllion, "PAVECHECK: Integration of Deflection and Ground Penetrating Radar Data for Pavement Evaluation," Texas DOT Report No. FHWA/TX-06/0-4495-1, Austin Texas, 2006.

43. Maser, K.R., and J. Puccinelli, "Ground Penetrating Radar (GPR) Analysis," Montana DOT Report No. FHWA/MT-09-005/8201, Helena MT, 2009.

44. Morey, R.M., "Ground Penetrating Radar for Evaluating Subsurface Conditions for Transportation Facilities," NCHRP Synthesis of Highway Practice 255, 1998.

45. Alavi, S., J.F. LeCates., and M.P. Tavares, "Falling Weigtht Deflectometer Usage, A Synthesis of Highway Practice," NCHRP Synthesis 381, Washington, D.C., 2008.

46. Dynatest, Falling Weight Deflectometer, http://www.dynatest.com/equipment/structural/fwd.aspx, accessed 2014.

47. Foundation Mechanics Inc., JILS, Equipment page, http://www.jilsfwd.com/product.html, accessed 2014.

48. Engineering & Research Intl. Inc., KUAB Falling Weight Deflectometers, http://www.erikuab.com/kuab.htm, accessed 2014.

49. Irwin, L.H., D.P. Orr, and D. Atkins, "Falling Weight Deflectometer Calibration Center and Operational Improvements: Redevelopment of the Calibration Protocol and Equipment," Report No. FHWA-HRT-07-040, 2011.

50. AMRL, "FWD Calibration Center Operator Certification Program," http://www.amrl.net/amrlsitefinity/default/fwd.aspx, accessed 2014.

51. Pavement Interactive, http://www.pavementinteractive.org/article/ backcalculation/, accessed 2014.

52. PCS Law Engineering, "Layer Moduli Backcalculation Procedure: Software Selection," SHRP-P-65 1, 1993.

53. Applied Research Associates, "Guide for Mechanistic-Empirical Design of New and Rehabilitated Pavement Structures," National Cooperative Highway Research Program Project 1-37A, Transportation Research Board, Washington, D.C., 2004.

54. Rhode, G. and Scullion, T., "MODULUS 4.0: A Microcomputer Based Procedure for Backcalculating Layer Moduli From FWD Data," Research Report 1123-1, Texas Transportation Institute, 1990.

55. Flintsch, G., S. Katicha, J. Bryce, B., Ferne. S. Nell, and B. Diefenderfer, "Assessment of Continuous Pavement Deflection Measuring Technologies," SHRP 2 Report S2-R06F-RW-1, Transportation Research Board, 2013.

56. Rada, G.R. and S. Nazarian, "The State-of-the-Technology of Moving Pavement Deflection Testing," Report FHWA-HIF-11-013, 2011.

57. Arora, J., V. Tandon, and S. Nazarian, "Continuous Deflection Testing of Highways at Traffic Speeds," Report No. FHWA/TX-06/0-4380-1, Center for Transportation Infrastructure Systems, University of Texas at El Paso, 2006.

58. Steele, D.A., and Vavrik, W. R. "Rolling Wheel Deflectometer (RWD) Results for the California Department of Transportation (Caltrans)," ARA Project No. 16860, Applied Research Associates, Champaign, IL, 2007.

59. Diefenderfer, B., "Investigation of the Rolling Wheel Deflectometer as a Network-Level Pavement Structural Evaluation Tool," Report No. FHWA/VTRC 10-R5, Virginia Transportation Research Council, Charlottesville, VA, 2010.

60. Elseifi, M., A.M. Abdel-Khalek, and D. Karthik, "Implementation of Rolling Wheel Deflectometer (RWD) in PMS and Pavement Preservation," Report No. FHWA A/11492, Louisiana State University, Baton Rouge, LA, 2012.

61. Elseifi, M.A., A. M. Abdel-Khalek; K. Gaspard; Z. Zhang, and S. Ismail, "Evaluation of Continuous Deflection Testing Using the Rolling Wheel Deflectometer in Louisiana," ASCE Journal of Transportation Engineering, Vol. 138, No. 4, April 1, 2012.

62. Hossain, M. and Chowdhury, T., "Use of Falling Weight Deflectometer Data for Assessing Pavement Structural Evaluation Values," Report No. K-Tran-KSU-96-4, Topeka KS, 1999.

63. Scullion, T., "Incorporating a Structural Strength Index into the Texas Pavement Evaluation System," Report No. FHWA/TX-88/409-3F, Texas Transportation Institute, College Station, TX, 1998.

64. Zhang, Z, Manuel, L., Damnjanovic, I, and Li, Z., "Development of a New Methodology for Characterizing Pavement Structural Condition for Network-Level Applications," Report No. FHWA/TX/-04/0-4322-1, Center for Transportation Research, Austin TX, 2003.

65. Bryce, J. Flintsch, G.W., Katicha, S.W., and Diefenderfer, B.K., "Developing a Network Level Structural Capacity Index for Structural Evaluation of Pavements," Report No. VCTIR 13-R9, Blacksburgh, Virginia, 2013.

66. Flora, W.F., "Development of a Structural Index for Pavement Management: An Exploratory Analysis," MSCE Thesis, Purdue University West, Lafayette IN, 2009.

67. Flora, W.F., G.P. Ong, and K.C. Sinha, "Development of a Structural Index as an Integtral Part of the Overall Pavement Quality in the INDOT PMS," Report No. FHWA/IN/JTRP-2010/11, Joint Transportation Research Program, West Lafayette, IN, 2010.

68. "Distress Identification Manual for the LTPP (Fourth Revised Edition)," Appendix B, *Manual for Faultmeter Measurements*, Publication Number: FHWA-RD-03-031, Washington, D.C., June 2003.

69. McGhee, K, "Automated Pavement Distress Collection Techniques," NCHRP Synthesis of Highway Practice 334: Automated Pavement Distress Collection Techniques, Transportation Research Board of the National Academies, Washington, D.C. 2004.

70. Wang, K.C.P., and O. Smadi, "Automated Imaging Technologies for Pavement Distress Surveys," Transportation Research Circular E-C156, Transportation Research Board, Washington, D.C., 2011.

71. Miller, J.S. and W.Y. Bellinger, "Distress Identification Manual for the Long-Term Pavement Performance Program (Fourth Revised Edition)," Report No. FHWA-RD-03-031, McLean VA, 2003.

72. Texas Research and Development Foundation, "Development and Implementation of Pavement Performance Data Collection Protocols," FHWA contract No. DTFH61-95-C-00019, Final Report, Austin, Texas, May 1997.

73. Sayers, M.W., "On the Calculation of International Roughness Index from Longitudinal Road Profile," the University of Michigan, Transportation Research Institute, Issue # 1501, Washington, D.C., http://www.trb.org/Publications/Pages/262.aspx,1995.

74. Surface Dynamics, Inc., "Smoothness Criteria For Runway Rehabilitation and Overlays," U.S. Department of Transportation, Federal Aviation Administration, National Technical Information Service, Report No. N91-19102, Springfield, Virginia, 1990.

75. Personal communication with Bob Olenoski, International Cybernetics, at Regional Profiling User Group (RPUG), September 1995.

76. The American Association of State Highway and Transportation Officials (AASHTO), *Pavement Management Guide*, 2nd Edition, Washington, D.C., 2012.

77. Underwood, B. S, A.M.; Y.R. Kim, and J. Corley-Lay, "Pavement Management Applications for Geographic Information Systems," ASCE Journal of Performance of Constructed Facilities, Vol. 25, No. 3, 2011.

78. Henning, T.F.P. and M.N.U. Mia, "Did We Get What We Wanted? – Getting Rid of Manual Condition Surveys," internet publication, accessed 4/16/2014 http://www.pavemetrics.com/pdf/3.CrackDetectionSurveysNewZealand.pdf.

79. Tsai, Y.C.J., and F. Li, "Critical Assessment of Detecting Asphalt Pavement Cracks under Different Lighting and Low Intensity Contrast Conditions Using Emerging 3D Laser Technology," "ASCE Journal of Transportation Engineering," Vol. 138, No. 5, 2012.

80. Wang, K.C.P., "Automated Survey of Pavement Distress based on 2D and 3D Laser Images," Report No. MBTC DOT 3023, Mack-Blackwell Rural Transportation Center, University of Arkansas, 2011.

81. Mergenmeier, A., "Pooled Fund Project: Improving the Quality of Pavement Surface Distress and Transverse Profile Data Collection and Analysis," Presentation, Road Profiler Users Meeting, San Antonio Texas, 2013.

82. NCHRP, "Automated Pavement Distress Collection Techniques," K. H. McGhee, *NCHRP Synthesis 334*, Transportation Research Board, Washington, D.C., 2004.

83. Papagiannakis, A., N. Gharaibeh, J. Weissmann, A. Wimsatt, "Pavement Scores Synthesis," Report No. FHWA/TX-09/0-6386-1, Texas Transportation Institute, Texas A&M University, College Station, Texas for the Texas Department of Transportation, 2009.

84. Louisiana Department of Transportation and Development, "Statewide Pavement Condition/Inventory Survey," *Contract for Special Services Project No. 736-099-1132*, LaDOT, Baton Rouge, Louisiana, October 2001.

85. Fugro Roadware, Inc. (USA), Austin, Texas, Information provided by Fugro to W.R. Hudson based on information available in their literature and on their website www.roadware.com/, 2014.

86. Pathway Service Inc., www.pathwayservices.com, Tulsa Oklahoma, 2010.

87. Permanent International Association of Road Congresses (PIARC), "International Experiment to Compare and Harmonize Skid Resistance and Texture Measurements," PIARC Ref. 01.04.TEN; ISBN 84-87825-96-6, 1995.

88. National Cooperative Highway Research Program (NCHRP), "Guide for Pavement Friction," Final Report for NCHRP Project 01-43, written by consultants J.W. Hall, K.L. Smith, L. Titus-Glover, Web Only Document, February 2009.

89. "Guideliness for Skid Resistant Pavement Design," AASHTO S99-SRPD-1, American Association of State Highway and Transportation Officials, Washington, D.C., 1976.

90. Titus-Glover, L., Owusu-Antwi, E.B., and Darter, M.I., *"Design and Construction of PCC Pavements, Volume III: Improved PCC Performance Models,"* FHWA-RD-98-113, Federal Highway Administration, Washington, D.C., January 1999.

91. Hall, J. P., "Challenges of Data for Performance Measures," Transportation Research Circular E-C115, Transportation Research Board of the National Academies, Washington, D.C., 2007.

92. Hoerner, T. E. and K. D. Smith, "High Performance Concrete Pavements: Pavement Texturing and Tire-Pavement Noise," FHWA-IF-02-020, Washington, D.C., 2002.

93. American Society for Testing and Materials, "Standard Smooth Tire for Pavement Skid-Resistance Tests," ASTM Standard Specification E-524, *Book of ASTM Standards*, Vol. 04.03, American Society for Testing and Materials, West Conshohocken, Pa., 1999.

94. American Society for Testing and Materials "Standard Practice for Computing International Roughness Index of Roads from Longitudinal Profile Measurements," ASTM E1926-08, 2008.

95. Transportation Association of Canada, "Pavement Asset Design and Management Guide," December, 2013.

96. American Public Works Association, "APWR Paver-Pavement Condition Index," APWA, 1984.

97. Karan, M.A., T.J. Christison, A. Cheetham and G. Berdahl, "Development and Implementation of Alberta's Pavement Information and Needs System," Transportation Research Board, Research Record 938, 1984.

98. Minnesota DOT (MnDOT), "An Overview of MnDOT's Pavement Condition Rating Procedure and Indices," website (www.MnDOT.org), May 9, 2006.

99. W.D.O. and T. Scullion, "Information Systems for Road Management: Draft Guidelines on System Design and Data Issues," Report INU77, The World Bank, Washington, D.C., 1990.

100. Flintsch, G. W., Dymond, R. and J Collura, "Pavement Management Applications for Geographic Information Systems," NCHRP 335, Transportation Research Board of the National Academies, Washington, D.C., 2004.

101. McPherson, K. and C.R. Bennett, *Success Factors for Road Management Systems*, The World Bank, Washington, D.C., 2005.

102. Schofer, J. L., T. Lomax, T. Palmerlee, and J. Zmud, "Transportation Information Assets and Impacts: An Assessment of Needs," Transportation Research Board Circular E-C109, Transportation Research Board of the National Academies, Washington, D.C., 2006.

103. Huene, K., "Data Sharing and Data Partnerships for Highways," NCHRP 288, Transportation Research Board of the National Academies, Washington, D.C., 2006.

104. Transportation Research Board, "Transportation Data and Performance Measurement," Summary of the Second National Conference on Performance Measures to Improve Transportation Systems, Washington, D.C., 2006.

105. Pickerell S. and L Neumann "Use of Performance Measures in Transportation Decision-Making," Transportation Research Board Conference on Performance Indicators, 2001.

106. TAC, "Measuring and Reporting Highway Asset Condition, Value And Performance," Report prepared by Stantec Consulting Ltd., Authors, Lynne Cowe Falls and Ralph Haas, 2001.

107. Haas, R. and W.R. Hudson, *Pavement Management Systems*, McGraw Hill, 1978.

108. International Asset Management Manual (IAMM) – v3.0. Thames, NZ: INGENIUM (Association of Local Government Engineering, NZ), Institute of Public Works Australia, Institution of Municipal Engineering Southern Africa, Institute of Asset Management UK, 2006.

109. Jarvis, Richard, "Developing Performance Measures for Asset Preservation in New South Wales," *Proceedings* 7th International Conference on Managing Pavement Assets, Calgary, June, 2008.

110. Jurgens, Roy and Jack Chan, "Highway Performance Measures for Business Plans in Alberta," *Proceedings* of the 2005 Annual Conference of the Transportation Association of Canada, Ottawa, 2005.

111. TAC, "Performance Measuring for Road Networks Guidelines," Report prepared by Stantec Consulting Ltd. and Ralph Haas, 2011.

112. Haas, Ralph, S. Tighe, J. Yeaman, L. Cowe Fall, "Long Term Warranty Provisions for Sustained Preservation of Pavement Networks," Paper prepared for presentation at the 2008 Annual Conference of the Transportation Association of Canada, Toronto, Canada , 2008.

113. Cowe Falls, L., R. Haas, S. Tighe, "Generic Protocol for Whole-of-Life Cost Analysis of Infrastructure Assets," *6th International Conference on Managing Pavements*, Queensland, Australia, 2004.

114. "Scanning the Home Front for Transportation's Best Practices," Washington, D.C., May-June Issue 2008 (See NCHRP Project 20-68, June 2007, for details).

Part Three

DETERMINING PRESENT AND FUTURE NEEDS AND PRIORITY PROGRAMING OF REHABILITATION AND MAINTENANCE

When a pavement is deficient in relation to some functional, condition, serviceability, or safety criteria, it becomes a "need for maintenance or rehabilitation." A "new need" refers to current deficiencies, and if a pavement is predicted to become deficient in some future year of the program period, it is identified as a future need. Because of budget restraints, few if any agencies can address all needs as they occur. That becomes the reason for priority programming of maintenance and rehabilitation, as described in the chapter.

14

Establishing Criteria

14.1   Reasons for Establishing Criteria

A criterion for establishing needs involves a specified limit on some pressure of performance, condition, behavior, safety, or other factor. The limit(s) should be objectively based and measurable, consistent, reasonable, and implementable.

14.2   Measures to which Criteria can be Applied

Criteria for possible needs or deficiencies are shown in the schematic of Figure 14.1. The criteria have limits in terms of maximum or minimum acceptability, either increasing or decreasing with time. Rather than establishing a single value for a limit, criteria may have a lower and upper bound, depending on the class of road, where action can be either desirable or absolute. Trigger values may also be set for criteria that pose high safety risks, such as inadequate skid resistances.

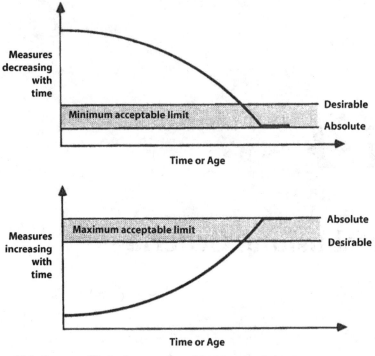

Figure 14.1 Concept of limits for measures of deterioration, behavior, response, or operating characteristics. After [1]

14.3   Factors Affecting Limits, and Some Examples

Factors that affect limits include the type and functional class of facility, size of pavement network, type of agency, and agency budgets and policies. Usually the type and class of facility dominate the other factors. The example criteria provided in Table 14.1 are not meant as recommendations but may be used for comparison and reference when establishing limits for a particular situation. It should be noted that actual limits for these criteria can vary considerably from agency to agency.

14.4   Effects of Changing Criteria

Changing criteria, in terms of which criteria to use, and/or changing the limits will advance or delay the needs years resulting in a decrease or increase of the amount of deficient mileage shown. This can affect the feasibility and cost-effectiveness of rehabilitation and maintenance treatments.

Table 14.1 Example limits for various measures. After [1]

Measure or Characteristic	Minimum or Maximum (Desirable) Acceptable Value				Remarks
	Freeway	Arterial	Collector	Local	
1. Roughness	Variable	Variable	Variable	Variable	Depends on how measured
a) PSI	3.0	2.5	1.5	1.5	
b) IRI	*	*	*	*	Remains to be established by IRI
c) RCI	6.0	5.0	4.0	3.0	
2. Surface Distress	Variable	Variable	Variable	Variable	Depends on distress type
a) SDI (scale of 0 to 10)	6.0	5.0	4.0	3.0	
b) PCI (scale of 0 to 10)	60	50	40	30	
3. Deflection	Variable	Variable	Variable	Variable	Depends on how measured
a) SAI (scale of 0 to 10)	7.0	6.0	5.0	4.0	
4. Surface friction	Variable	Variable	Variable	Variable	Depends on how measured
a) Skid number (ASTM)	*	*	*	*	Not specified by hwy agencies
5. Combined index					
a) PQI (scale of 0 to 10)	6.0	5.0	4.0	3.0	
6. Traffic delays (Veh. hours)	*	*	*	*	Remains to be established
7. Vehicle operating costs	*	*	*	*	Remains to be established

15

Prediction Models for Pavement Deterioration

Predictions of pavement performance or definition are essential to establish needs. In turn, predictions for maintenance and rehabilitation treatment alternatives are essential for priority programming.

15.1 Clarification of Performance and Deterioration Prediction

The challenge of developing good models for predicting performance in terms of a measure such as Present Serviceability Index (PSI) or Riding Comfort Index (RCI) versus age or accumulative axle load applications has existed since the advent of pavement management. It has become common among practitioners to interchange the word performance with alternate terms such as deterioration or damage. In itself, this is not a problem as long as the specific measure involved is identified.

15.2   Parameters or Measures to be Predicted

In order to estimate future needs years for sections in a pavement network, it is important to predict the rate of change of the performance measures used. Figure 15.1 [1] illustrates how deterioration prediction would be applied to an existing pavement section to estimate the rate of future deterioration and rehabilitation alternatives. The basic requirements for any prediction model are represented in the figure.

15.2.1   Deterioration Prediction Model Approaches and Variables

Deterioration or performance prediction is essential to life cycle cost analysis of road infrastructure. A variety of models and approaches exist that can be applied at the following levels.

- Strategic level: to identify long range needs.
- Network or system wide level: where performance estimates are made for new designs, rehabilitation, or maintenance options on the network.
- Project or site specific level: where more detailed deterioration functions are applied.

The strategic level usually involves remaining service life estimates of the road infrastructure component (e.g., future years of overlay, reconstruction,

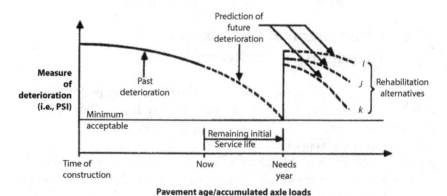

Figure 15.1 Illustration of how a deterioration model is used to predict future deterioration of an existing pavement, and rehabilitation alternative constructed in the "needs" year. After [1]

etc., which, when compared to what actually can be funded, enables identification of the infrastructure "backlog" or "gap"). Network or system deterioration models are usually developed and applied to a grouping of components or subgroups within a component (e.g., same model for all pavements with similar structural number or granular base equivalency range). At the project or site specific level, the deterioration models can range from approximate to sophisticated depending largely on the road component and knowledge about it.

Of the many technical requirements for effective road asset management, the most important is a combined structural analysis and performance prediction methodology [2]. This is commonly referred to as mechanistic-empirical (ME), where the mechanistic part is used to calculate one or more responses in the pavement structure as a function of material properties, layer thicknesses, loading conditions, temperature, etc. These responses must then be related to observed performance; e.g., smoothness deterioration, fatigue cracking progression, and rutting progression. That is the empirical part of ME design. Figure 15.2 shows a typical approach; however, establishing the combined relationship is a real challenge to pavement engineers.

A summary of computer-based analytical solutions for mechanistic analysis is provided in [3] and updated in [2]. The basic form of deterioration consists of one or more of the following elements: straight line, concave up, concave down, or a step function where sudden damage occurs. In Figure 15.3 the dotted line distributions represent variance, the deterioration can be both concave up and concave down, and there is a minimum acceptable level (often called a "trigger level") at which point an intervention should be carried out, depending on available funding and priority need [2].

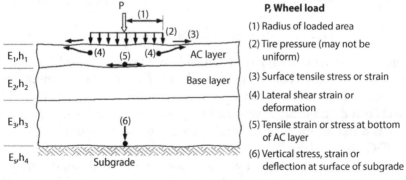

Figure 15.2 Fundamental pavement responses as a function of load, material properties and layer thicknesses (mechanistic part).

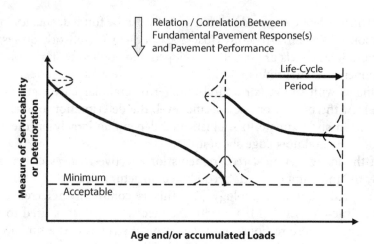

Figure 15.3 Pavement performance to which mechanistic response(s) must be related/correlated (empirical part). After [2]

One of the biggest challenges in deterioration modeling is to identify and characterize the independent variables which affect deterioration. A complementary challenge is to balance the level of effort, time, and cost of obtaining the data to arrive at a reasonably reliable prediction in a reasonable timeframe. To illustrate this challenge, Figure 15.4 shows the groups of variables or factors and the component factors which can affect pavement deterioration. Also shown, by dotted lines, are the interactions of factors that can occur [2] .

While it would be desirable, ideally, to quantitatively characterize all the factors, sub factors, and interactions, this is obviously impractical. Thus, most working models have incorporated aggregated factors that are calibrated to local or regional conditions, and they often use surrogates as independent variables (e.g., age to represent environment or climate influence).

15.2.1.1   Project Level/Site Specific Deterioration Modelling

Project level deterioration models range from purely empirical (e.g. based on regression analysis of observed data) to fairly sophisticated mechanistic-empirical, such as developed in NCHRP Project 37-A, the AASHTO "Mechanistic-Empirical Pavement Design Guide" (MEPDG) [4]. While there is a trend toward more fundamentally based models, like those in MEPDG, these also require calibration to local or regional conditions.

A summary of the 12 most widely known and/or used and/or internationally recognized design procedures which incorporate mechanistically based deterioration models, including the MEPDG, is provided in [5].

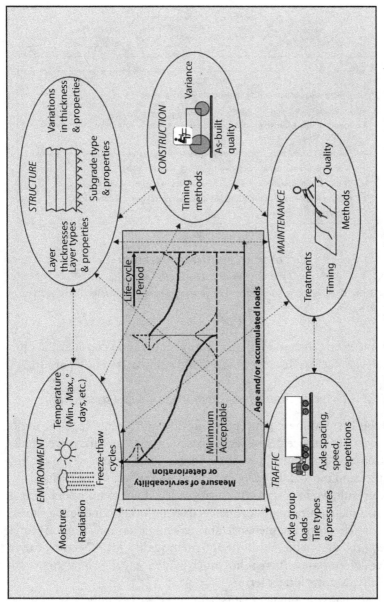

Figure 15.4 Factors, sub factors and interactions which can affect pavement deterioration. After [2]

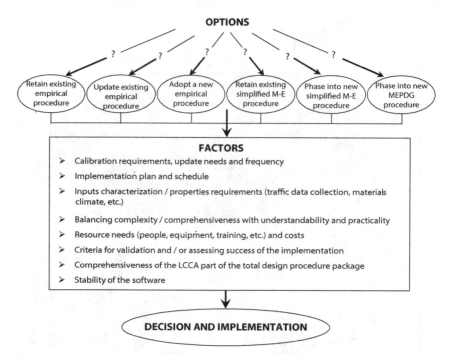

Figure 15.5 Basic options and factors in choosing a pavement design deterioration model. After [5]

In adopting any such model, the road agency should consider the options and relevant factors discussed by [5], and summarized in Figure 15.5.

15.2.1.2   *Deterioration Modelling for Long Life Pavements*

The long life pavement concept for flexible pavements requires periodic monitoring of surface distresses [6] based on a greater likelihood of deterioration occurring in the wearing course than deeper in the pavement structure and the fact that deeper failures also reflect to the surface.

As a result, when design criteria are satisfied, for example reaching limiting cumulative strain, performance, or deterioration modeling can show need for scheduled maintenance and rehabilitation interventions to yield the required design life. While design methods like MEPDG can be used to predict deterioration, there is not much evidence to date on their accuracy, especially over the longer term.

An alternative approach that is being used successfully in several states, provinces, and cities (e.g., Minnesota, Ontario, and others) involves individual section monitoring and performance prediction using an auto adaptive, sigmoidal model. This approach responds to activities performed on the pavement, rather than type, and to environment, traffic, and ongoing maintenance effects. The form of the model is:

$$P = P_0 \pm e^{(a-b^*c^t)}$$

where P = performance measure (e.g., PCI, where a subtraction is made; or IRI, where addition is made).

P_0 = P at age 0
$t = \log_e (1/age)$
a, b & c = model coefficients

Depending on the model coefficients, the relationship can be a straight line, concave, convex or S shaped, with varying degrees of curvature. The auto adaptation procedure consists of calculating the model coefficients through non-linear regression each time a new IRI/PCI, etc. is determined for the pavement section. In essence, this site specific modelling approach enables tracking of past performance to the current year of measurement and prediction of future preformance. The ability to closely track past performance and predict future performance has been illistrated for I35 in Minnesota, South Carolina, and elsewhere by [7]. Auto-adaptation is built into the system so that each time a new IRI/PCI is uploaded to the data base, the model coefficients are recalculated.

Individual site specific distresses can also be predicted using the following sigmoidal model form:

$$D = e - (k / age)^b$$

where D = distress density (0 to 1, where 1 would be 100% of the area)
   Age = years since the last rehabilitation
   K, b = model coefficients specific to distress type and severity

In this formulation, limits can also be defined on service life to constrain the coefficients to produce reasonable models. An example use of this approach took place in Minnesota [7].

15.3   Basic Types of Prediction Models and Examples

There are four basic types of prediction models: 1) purely mechanistic, 2) mechanistic-empirical, 3) regression based, and 4) subjective. While the examples given in [1] are valuable for illustration, a large amount of development in recent years has gone into M-E in the AASHTO's Mechanical Empirical Pavement Design Guide, as discussed in Part Four. A summary of the prediction approach, rather than the details of a complex modeling process, is provided below.

15.3.1   Performance Prediction Approach in the Mechanistic Empirical Pavement Design Guide (MEPDG)

The MEPDG was originally developed in NCHRP Project 1-37A [8], 2004. Since then an enormous amount of federal, state/provincial, and local authority's efforts have gone into adaption/implementation and calibration (see Part Four, "Project Level Design: Structural and Economic Analysis" for a description of the MEPDG). The software package was originally named DARWin-ME, which as of 2013 is called Pavement ME [9]. While the MEPDG has been calibrated nationally, the AASHTO implementation guide recommends local calibration of the models [4].

Since the MEPDG incorporates over 350 variables, including environment/climate, traffic load spectra, materials characterization that considers ageing throughout the design life, drainage, foundation, and others, the sensitivity of these variables, particularly in combination for any given design situation, is a huge issue. Obtaining real data is another issue. Unlike the global sensitivity evaluations provided as part of NCHRP Project 1-47 [10], MEPDG users have often tried to evaluate sensitivity of variables one, two, or three at a time. Such an approach can be misleading since interacting effects of variables are not captured this way.

To predict performance, MEPDG uses a specific set of inputs selected for a trial design. Performance and distress are calculated and compared to specific criteria at the end of a specified design life for a specified level of reliability. For flexible pavements, these criteria would normally include roughness, distress in terms of bottom up and top down fatigue cracking, thermal cracking, and pavement deterioration. The trial designs that meet the criteria are compared in a life cycle economic analysis to identify the one that is most cost-effective.

A project carried out in Canada, using Pavement ME, evaluated typical flexible pavement designs at different weather/climate stations with varying thickness, truck traffic levels, and different levels of Performance Graded Asphalts [11]. While not a global sensitivity analysis, the authors found the results to be reasonable and valuable in ways to obtain familiarity with the design package.

16

Determining Needs

One approach to determine maintenance and rehabilitation needs is to highlight "needs" years, which may or may not be "action" years, and their relation to prediction models. For example, in Figure 15.1, the needs year for the rehabilitation alternatives is also the action year. But action years can be deferred or advanced from "needs" years according to the available resources or priority alternatives. "Needs" and "action" years can also vary when minimum acceptable deterioration levels are changed; however, there are practical and economic limits to the range of "action" years.

"Needs" and possible "action" years do not indicate what maintenance or rehabilitation alternatives should be considered. For small networks, rehabilitation alternatives can be based on the experience and judgment of the local engineer, whereas large networks require a priority programming approach, as discussed in Chapter 18. In that type of approach, alternative action year and alternative maintenance and rehabilitation treatment combinations can be evaluated in searching for an optimum combination.

17

Rehabilitation and Maintenance Alternatives

Feasible maintenance and rehabilitation alternatives and targets are identified using deterioration modeling for cost and applicability analysis. Various rehabilitation and maintenance costs and benefits are involved along with applicable technologies, calculations, and models. Many of these are appropriate, as described in the following sections.

17.1 Identification of Alternatives

The alternatives considered by an agency usually represent current practice but can change based on new technology assessment by other agencies and observed long-term performance. Identifying alternative rehabilitation strategies provides for altering existing strategies and procedure for deciding which ones are feasible for a given situation in order to efficiently establish program priorities. For example, Ontario identifies alternatives for flexible and rigid pavements in Figure 17.1 classified as Rehabilitation, Routine Maintenance, and Major Maintenance. Maintenance treatments

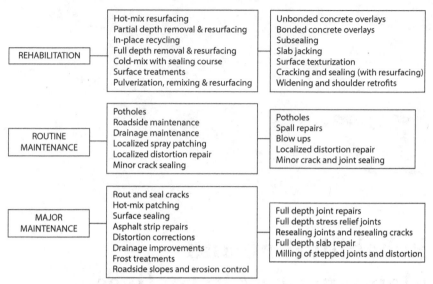

Figure 17.1 Rehabilitation and maintenance alternatives used in Ontario. After [12].

can be classified as preventive or corrective. Currently, the term "Pavement Preservation" essentially replaces or incorporates "Preventive Maintenance," as discussed in the following section.

17.1.1   Pavement Preservation

The concept of pavement preservation has been largely championed by the US Federal Highway Administration and its web site is replete with references, terminology, etc. (www.thwa.dot.gov/pavement/pres.cfm). Several other agencies have been active in the concept, including The National Center for Pavement Preservation at Michigan State University (www.pavementpreservation.org), The California Pavement Preservation Center at California State University, Chico (www.csuchico.edu/cp2c/), and the Texas Pavement Preservation Center in collaboration with The Center for Transportation Research of The University of Texas, Austin, and The Texas Transportation Institute of Texas A&M University. There is also a "Pavement Preservation Journal" (www.fp2.org), which is actually a promotional forum, and there is a lot of information on example projects, preservation technologies, evaluations, and conferences on the above web sites.

The most common or widely used preservation treatments for flexible pavements vary with reference source but generally include the following:

- Microsurfacing, slurry seals and fog seals
- Crack sealing, spray patching, and full depth patching
- Thin overlays and milling plus overlay ("mill and fill")
- Chip seals/seal coats

For rigid pavements, common preservation techniques include the following:

- Diamond grinding
- Full or partial depth slab repairs
- Dowel bar/load transfer retrofits
- Joint and crack sealing
- Hot mix asphalt overlay

17.1.2 Examples of Combined Rehabilitation and Preventive/Preservation Treatment Alternatives at the Network Level

The ICMPA7 Investment Analysis and Communication Challenge for Road Assets [13] involved a network of 1,293 road sections, 161 bridges, 356 culverts, and 45 major signs. Part of the data base, models, costs, etc. provided for the challenge was a set of rehabilitation and preventive maintenance treatments, given in Table 17.1, where the latter might also be called preservation treatments. Unit costs, expected service lives, and other information are also provided. Guidelines for selecting rehabilitation and maintenance treatments from the alternatives in Table 17.1 are given in the decision tree, Figure 17.2.

17.2 Decision Processes and Expert Systems Approaches to Identifying Feasible Alternative

The process of selecting feasible rehabilitation alternatives can range from simple judgment to the use of a decision tree. An example decision tree for rigid pavement rehabilitation in Figure 17.3 from [14] is still relatively applicable.

The use of Expert Systems to identify feasible alternatives was quite promising in the 1990s. While the approach is still valid, a simpler approach involving a decision matrix is currently more commonly used.

Table 17.1 Pavement rehabilitation and preventive maintenance treatments. After [13]

No.	Treatment	Type¹	Applicability	Unit Costs²	Expected Service Life	Expected Effect	Remarks
1	Thin Overlay (40 mm or less in thickness)	PM	Rough pavements with or without surface deficiencies but structurally adequate; can be applied to structurally inadequate pavements to defer grade widening or reconstruction. Would not generally be considered for high volume roadways	$6.00/m² to $7.50/m²	Structurally adequate pavement: ≤ 10 years / Structurally inadequate pavement: ≤ 5 years	Reduces IRI	• Can treat travel lanes only or full width • May not be able to meet QA smoothness specifications
2	Reprofiling by Cold Milling and Overlay	SP	Rough pavements with or without surface deficiencies and modest strengthening needs	$9.00/m²	≤ 15 years	Reduces IRI and improves general roughness; restores structural integrity	• Overlay based on structural design
3	Cold Mill and Inlay, or HIR of Travel Lanes, plus Overlay	SP	Pavements with severe surface deficiencies and strengthening needs as determined by condition evaluation and/ or deflection testing or other means	$15.00/m² to $16.50/m²	≤ 15 years	Reduces IRI and improves general roughness; restores structural integrity	• Overlay based on structural design
4	Structural Overlay	SP	Structurally deficient pavements as determined by condition evaluation and/or deflection testing, or other means	$10.50/m² to $16.50/m²	10 year design: 10 years / 20 year design: 20 years	Reduces IRI and improves general roughness, increases or restores structural integrity	• Structural deficiency can result from under-design or increased traffic loading • Overlay thickness based on structural design

5	Cold Mill and Inlay	PM	Rough and/or rutting distress but structurally adequate pavements; interim measure to improve ride quality until overlay needed	$9.00/m²	Structurally adequate pavement: ≤ 10 years	Reduces IRI and improves surface condition	• Typically 50 mm cold mill depth • Treatment applied to travel lanes only
					Structurally inadequate pavement: < 10 years		
6	Hot-In-Place Recycling (HIR)	PM	Rough but structurally adequate pavements; interim measure to improve ride quality until overlay needed	$7.50/m²	Structurally adequate pavement: ≤ 8 years	Reduces IRI and improves surface condition	• Pavements with severe deficiencies (e.g. rutting) may not be suitable candidates • Seal Coats, patching and crack sealer may affect mix quality • Treatment applied to travel lanes only
					Structurally inadequate pavement: < 8 years		
7	Micro-Surfacing	PM	Structurally sound, relatively smooth pavements which may have some surface distress (e.g. raveling, segregation); can also be used as a rut fill treatment	$4.50/m² to $6.00/m²	5 years	Seals surface and may increase surface friction	• May be appropriate for semi-urban applications

(Continued)

Table 17.1 Pavement rehabilitation and preventive maintenance treatments. After [13] (*Continued*)

No.	Treatment	Type[1]	Applicability	Unit Costs[2]	Expected Service Life	Expected Effect	Remarks
8	Chip Seal (Surface Seal; Seal Coat)	PM	Structurally sound, relatively smooth pavements; may have some surface distress (e.g. raveling)	$3.75/m^2	≤ 7 years	Improved surface friction; extended service life of pavement	• No added structural strength
9	Cold In-Place Recycling	RC	Pavements for which preventive maintenance or rehabilitation is not an option (e.g. excessive roughness and/or structural damage)	≈$37.50/m^2	≤ 20 years	Restores IRI, restores structural integrity	• Need surface wearing course
10	Full Depth Reclamation and Stabilization	RC	Pavements for which preventive maintenance or rehabilitation is not an option (e.g. excessive roughness and/or structural damage)	≈$37.50/m^2	≤ 20 years	Restores IRI, restores structural integrity	• Need surface wearing course
11	Reconstruction	RC	Pavements for which preventive maintenance or rehabilitation is not an option (e.g. excessive roughness and/or structural damage)	≈$37.50/m^2	20 years	Restores IRI, restores structural integrity	• Replaces existing structure

[1]PM is preventive maintenance; SP is strengthening (e.g. structural preservation); RC is reconstruction; [2]Expected 2008 unit costs

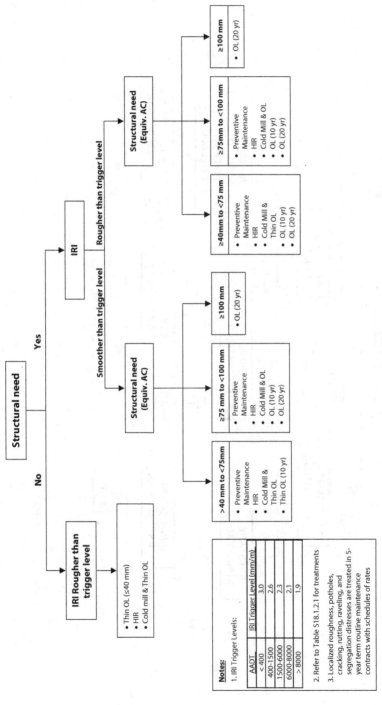

Figure 17.2 Guidelines for selecting pavement rehabilitation and preventive maintenance treatments. After [13]

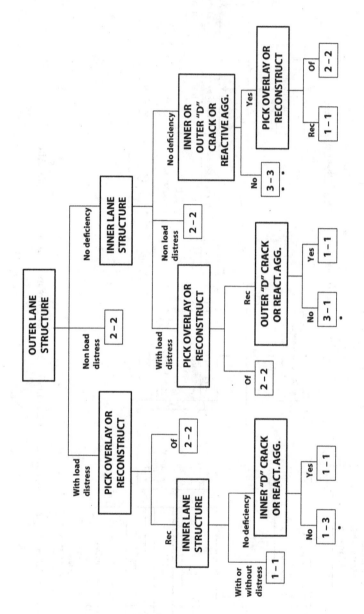

1 – 1 ReconstructBoth Lanes
1 – 3 Reconstruct Outer, Restore Inner
3 – 1 Restore Outer, Reconstruct Inner
3 – 3 Restore Both Lanes
2 – 2 Overlay Both Lanes

• Option to Go to 1 – 1 Provided
•• Option to Go to 1 – 1, 1 – 3, or 2 – 2 Provided

Figure 17.3 Decision tree for selection of main rehabilitation approach. After [14].

17.3 Deterioration Modeling of Rehabilitation and Maintenance Alternatives

Many types of models are applicable to modeling the deterioration of rehabilitation alternatives. For example, deterioration modeling in AASHTO's Mechanistic Empirical Pavement Design Guide (MEPDG) uses the same approach for rehabilitation treatments as for new pavement designs [4].

Another simpler approach is to assign expected service life to routine maintenance, rehabilitation, and pavement preservation treatments as given in [15]. Deterioration modeling in the network of 1,293 road sections of the "Challenge" [13] was estimated by linear regression and ranged from an IRI increase of 0.069 m/km/year to 0.101 m/km/year, depending on road class, traffic volume, and rehabilitation treatment.

17.4 Costs, Benefits, and Cost-Effectiveness Calculations

Calculations for costs, cost-effectiveness, and benefits for rehabilitation alternatives are needed for priority programing. Various rehabilitation costs, including cost of actual work, vehicle-operating costs, cost of user delay, etc., are identified and described in [1], as are relevant formulas and example calculations. These approaches and methods are still relevant, but the actual numbers would have to be scaled up for current conditions and cost levels. Costs and benefit calculations are fully discussed in Part Seven.

18

Priority Programing of Rehabilitation and Maintenance

All agencies must establish a basic approach and policy for their rehabilitation and maintenance programs, including the selection of appropriate program periods. The basic functions of priority programming, various methods and examples, are provided in [1]. Budget level evaluation in terms of the effects of different budget levels is also described, as is the effect of specified standards.

18.1 Basic Approaches to Establishing Alternatives and Policies

Basic approaches used by agencies to establish alternatives and policies for their rehabilitation and maintenance programs include the strategic approach, defining a set of approved and a limited number of alternatives. The strategic approach of setting targets is particularly important to good planning and policy. As well, an approved set of alternatives periodically updated is a useful way to define those that are practical, cost effective, regionally applicable, and constructible with available resources and materials.

18.2   Selecting a Length of Program Period

While a pavement life-cycle economic analysis might be 20 years or more, a program period for maintenance work would generally be only one to five years with annual and biannual updates. A practical 5- or 10-year program for rehabilitation is often set, in which the first two or three years are fixed (for detailed design and contracting reasons) but annual or biannual updates are also used. Modern PMS or MMS software will contain such models and help formulate such programs.

18.3   Basic Functions of Priority Programming

Comparing investment alternatives at either or both the network and project level should result in a priority program of new pavement construction, and/or rehabilitation, and/or maintenance. However, use at the project level is a small subset of the network and the overall network budget is overriding. Any priority program should respond to What, When, and How. Although not simple to do, all possible combinations of actions should be evaluated. The major steps in priority programing (Figure 18.1) show how to apply the process.

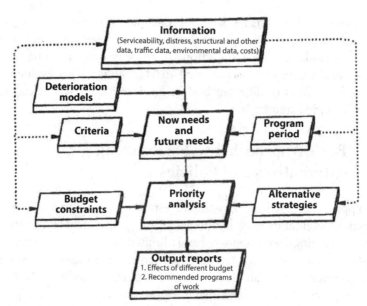

Figure 18.1 Major steps in priority programming. After [1]

18.4   Priority Programing Methods

Priorities ranging from simple subjective ranking to comprehensive optimization are summarized in Table 18.1 [16]. The use of genetic algorithms in optimization saw considerable applications in the 1990s, as subsequently discussed. Neural networks have also been used at the project level but seldom as a network optimization or priority programming tool.

18.4.1   Mathematical Programming for Optimization Method

Mathematical programming methods are based on formulating an objective function to maximized or minimized some specified value such as costs or benefits, which are in the case of costs subject to constraints such as budgets. The program period involved can be single year or multi year. Budget constraints may cause projects to be delayed. When that happens user benefits are reduced as schematically shown in Figure 18.2 [16].

Table 18.1 Different classes of priority programming methods. After [16].

Class of Method	Advantages and Disadvantages
Simple subjective ranging of projects based on judgment	Quick, simple; subject to bias and inconsistency; may be far from optimal
Ranking based on parameters, such as serviceability, deflection, etc.	Simple and easy to use; may be far from optimal
Ranking based on parameters with economic analysis	Reasonably simple; should be closer to optimal
Optimization by mathematical programming model for year-by-year basis	Less simple; may be close to optimal effects of timing not considered
Near optimization using heuristics and marginal cost-effectiveness	Reasonably simple; can be used in a microcomputer environment; close to optimal results
Comprehensive optimization by mathematical programming model taking into account the effects	Most complex; can give optimal program (max. of benefits)

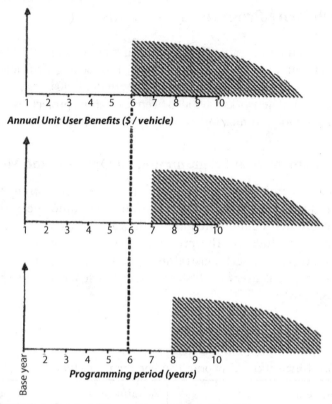

Figure 18.2 Effect of project delays on user benefits. After [16]

18.4.2   Genetic Algorithms and Evolutionary Algorithms as an Optimization Tool

The principals of evolution in nature form the basis of genetic algorithms by a procedure that encodes a series of bit strings. The algorithm then manipulates these strings in part by random sampling and random changes called "mutations" to arrive at the "most fit" members of the population involved. There are many example applications in the literature, such as an application on road maintenance planning [17], then an application on multi objective optimization in pavement management [18], and later a maintenance strategy optimization for rural road networks [19]. Microsoft Excel solver is a genetic algorithm available from "Frontline Systems" (www.solver.com/generic-evolutionary-introduction). Another generic algorithm tool is Evolver, Version 32.0, available from the Palisade Corporation (www.palisade.com). It can be used as an add-on for Microsoft Excel.

Linear programming is used in the optimization. Evolutionary Algorithms, also available from Frontline Systems, use the decision variables, problem objective functions, and constraints directly. These developers contend that great use is now made of the Evolutionary Algorithm solver products. Be cautioned that any solutions arrived at are not necessarily optimal; they are only better in comparison to other tested solutions. The user also needs to specify the number of iterations or candidate solutions because the algorithm does not otherwise know when to stop.

18.4.3   Neural Networks as an Optimization Tool

Neural networks have been used in civil engineering, especially in complex problems difficult to model or solve by conventional approaches [20]. Essentially, they work as a learning process and can be "trained" through a single error back progression method. A Google search, using "neural networks in pavement management," provides many hits on pavement project level applications for distress prediction, roughness prediction, and other uses.

There has been application of neutral networks in optimal signal timing at intersections, for example, in bridge management for budget allocation [20]. However, unless combined with other models, neural networks have found little application as an optimization tool in pavement management.

18.5   Examples and Comparisons

Comparison of network priority programing methods were not readily available in the 1990s. Essentially that is still the situation in 2014. It is well known that different methods can produce different priority program results; therefore, users should carefully vet the type of optimization used in any software considered for use.

18.6   Budget Level Evaluation and Specific Standards

Agencies need a tool to assess the effects of different budget or funding levels on overall network quality (serviceability). Figure 18.3 [21] plots average pavement quality (scale of 0 to 10) versus time for a pavement network and shows how much average quality changes for various budget levels. This gives decision makers a clearer picture of the cumulative effect of low budgets. Corresponding examples of accumulation of deficient mileage or backlog can be developed.

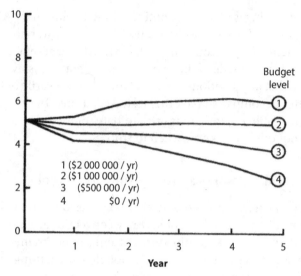

Figure 18.3 Budget level evaluation for the example network. After [21].

With good PMS software, the agency can also specify a desired average performance standard and use the PMS to estimate funds needed to meet that standard. This is simple for a year-to-year horizon but difficult for multiyear horizons. A very limited number of software providers known to the authors can do true multiyear optimization.

18.7   Final Program Selection

Any good PMS software will provide the agency decision makers with several levels of summary information: for example, answers with and without consideration of user costs, single year or multiyear, etc. But software cannot make a final decision because it cannot fully access qualitative information such as political climate, taxation, rural versus urban concerns, etc. Remember, the final decision is up to the agency to make based on the software output. Furthermore, parameters can be adjusted, as can new scenarios, in making final decisions.

The final decision may differ from the mathematical optimum. The good news is that you input the resulting decisions and actions into the PMS data base and the next set of optimization take them into account in future calculations. If the changes are drastic, the decision makers can use the negative results to show legislators or funding agencies the need for more or less funding. This is an additional benefit of good "feedback."

19

Developing Combined Programs of Maintenance and Rehabilitation

Sometimes it is desirable to combine maintenance and rehabilitation programs. It can be useful to combine optimization for preventive and/or preservation maintenance, rehabilitation, and specified pre-and post-rehabilitation maintenance for a life-cycle. Such an approach should result in the best overall use of resources. One of the main barriers is institutional since many agencies have separate budgets for rehabilitation and maintenance.

Pre-rehabilitation involves needed section maintenance when planned rehabilitation is delayed. Figure 19.1 illustrates an example of pre-rehabilitation strategies and points out the maintenance cost portion can be calculated with unit costs assigned to each maintenance treatment [22]. This is termed preventive maintenance or pavement preservation. Similar calculations can be made for post rehabilitation strategies for overall life cycle economic analysis.

IMPLEMENTATION YEAR	MAINTENANCE STRATEGY & PERCENT OF SECTION FOR YEARS PRIOR TO IMPLEMENTATION										NO.OF MAINT. YRS FOR REMOVAL (1)	NO.OF MAINT. YRS FOR RECONSTR (2)
	1	2	3	4	5	6	7	8	9	10		
1	1 - 80										1	1
2	5 - 20	1 - 70									2	2
3	1 - 40	3 - 5	1 - 60								2	2
4	2 - 7	3 - 5	5 - 20	1 - 50							2	3
5	1 - 60	2 - 8	3 - 5	5 - 20	1 - 40						2	3
6	1 - 50	3 - 3	2 - 8	3 - 7	5 - 20	1 - 40					2	3
7	1 - 25	1 - 25	4 - 20	2 - 7	3 - 8	5 - 25	1 - 40				2	4
8	1 - 10	1 - 25	4 - 20	4 - 20	5 - 20	3 - 10	5 - 20	1 - 40			2	4
9	1 - 5	1 - 15	5 - 15	2 - 5	4 - 20	2 - 7	5 - 20	3 - 15	1 - 40		3	5
10	1 - 5	1 - 15	5 - 15	2 - 3	5 - 15	4 - 20	5 - 15	2 - 5	1 - 30	5 - 30	0	0

Maintenance Strategies
1. Crack seal
2. Hot mix patch, 2 in.
3. Cold mix patch, 2 in.
4. Slurry seal
5. Fog seal
6. Fabric+ Thin overlay

(1) If the implementation alternative is a removal, such as recycling or milling, maintenance is eliminated for the number of years specified immediately prior to the implementation year.

(2) If the implementation alternative is a reconstruction, maintenance is eliminated in the number of years specified immediately prior to the implementation year.

Figure 19.1 Example of pre-rehabilitation maintenance strategies. After [22]

19.1   Example Results of a Combined Program

An example of a combined strategy including the costs and budget requirements can demonstrate the feasibility of developing cohesive pavement strategies. However, overcoming institutional barriers is a problem.

19.1.1   Example Results of a Combined Program Using the World Bank's HDM-4 Model

The World Bank's Highway Development and Management Model (HDM-4), a suite of applications and analysis tools (HDM-4 Version 2.05) (http://go.worldbank.org/JGIHXVL460), was used by [23] to illustrate a combined program for 1,293 road sections, 161 bridges, 356 culverts, and 45 major signs in the "Challenge" [13]. Chamorro first defined a representative road network for modeling in HDM-4, then characterized the vehicle fleet, selected maintenance standards (structural and preventive), applied vehicle operating and maintenance treatment unit costs, and defined other required inputs. The system was optimized to minimize the net present value of costs for a target roughness level.

While the analysis was for a 20-year period, a combined program example for the first five years is given in Table 19.1. It is noteworthy that many of the treatments are prevention or preservation.

A summary of the combined program effect on roughness for three funding levels is shown in Figure 19.2. The average IRI decreases with increased funding. The relative percentages of structural and preventive maintenance are provided in Figure 19.3. In essence, this example, which is described in detail in [23], illustrates the value of overall optimization for a combined program.

19.2   Summary

It is beyond the scope of this book to cover the methods of priority programming and combined programs in detail, including approaches used in other countries. Those who desire more information on this type of programming should refer to a specialist. Any good commercial PMS software provider should be able to describe details. As well, the International Conferences on Managing Pavement Assets (ICMPA's), with eight meetings held to date and ICMPA9 scheduled for Alexandria, Virginia, in 2015, represent a valuable repository of North American and international information on the subject.

Table 19.1 Example results of a combined program. After [23]

Year	Section Code	Road Class	Length (km)	AADT	Work Description	Treatment Type	Financial Costs (CAD$M)	Cumulative Costs (CAD$M)
2008	R-H-F-6A	Rural	10,90	6711	Cold Mill and Inlay 60mm	Preventive	1,18	1,18
	R-L-F-150C	Rural	36,94	229	Thin Overlay 30 mm	Preventive	1,99	3,17
	R-L-P-135I	Rural	25,10	292	Cold Mill and Inlay 70mm	Structural	2,37	5,54
	R-L-P-141B	Rural	30,83	386	Thin Overlay 30 mm	Preventive	1,29	6,84
	R-L-P-231B	Rural	32,68	259	Thin Overlay 30 mm	Preventive	1,76	8,60
	R-L-P-96A	Rural	64,77	410	Thin Overlay 30 mm	Preventive	3,50	12,10
	R-L-P-96B	Rural	44,15	172	Thin Overlay 30 mm	Preventive	2,38	14,49
	R-VH-P-3A	Rural	10,90	10860	Cold Mill and Inlay 60mm	Preventive	0,88	15,37
	R-VH-P-6A	Rural	5,09	11205	Cold Mill and Inlay 60mm	Preventive	0,41	15,78
2009	I-L-P-75B	Interurban	100,26	1499	Reseal at 5% surface crack	Preventive	4,51	20,29
	R-L-G-138A	Rural	39,35	606	Cold Mill and Inlay 70mm	Structural	4,96	25,25
	R-L-G-237A	Rural	39,41	149	Reseal at 5% surface crack	Preventive	1,33	26,58
	R-L-P-6D	Rural	11,28	358	Thin Overlay 30 mm	Preventive	0,61	27,19
	R-L-P-96C	Rural	24,30	216	Thin Overlay 30 mm	Preventive	1,31	28,50
	R-L-P-99A	Rural	22,78	139	Thin Overlay 30 mm	Preventive	1,23	29,73

Year								
2010	R-L-F-138C	Rural	45,55	396	Reseal at 5% surface crack	Preventive	2,05	38,24
	R-L-F-138B	Rural	51,24	498	Cold Mill and Inlay 70mm	Structural	6,46	36,19
	R-L-G-102B	Rural	38,47	555	Cold Mill and Inlay 70mm	Structural	4,85	43,08
	R-L-G-195A	Rural	44,03	235	Cold Mill and Inlay 70mm	Structural	4,16	47,25
2011	R-L-G-237D	Rural	45,93	179	Reseal at 5% surface crack	Preventive	1,21	48,45
	I-L-P-75B	Interurban	100,26	1654	Thin Overlay 40 mm	Preventive	8,42	56,87
	I-M-F-75G	Interurban	72,77	4254	Reseal at 5% surface crack	Preventive	3,27	60,15
	R-L-F-132C	Rural	35,57	334	Thin Overlay 30 mm	Preventive	2,56	62,71
2012	R-L-F-237C	Rural	29,15	142	Thin Overlay 30 mm	Preventive	1,22	63,93
	R-L-F-6G	Rural	27,09	144	Cold Mill and Inlay 70mm	Structural	2,56	66,49
	R-L-G-135H	Rural	33,49	435	Cold Mill and Inlay 70mm	Structural	4,22	70,71
	R-L-P-132D	Rural	40,45	195	Thin Overlay 30 mm	Preventive	2,18	72,90
	R-L-P-231A	Rural	37,71	273	Thin Overlay 30 mm	Preventive	2,04	74,93
	I-L-P-72B	Interurban	36,98	1231	Reseal at 5% surface crack	Preventive	1,66	76,60
2013	I-M-G-75F	Interurban	55,86	4165	Reseal at 5% surface crack	Preventive	2,51	79,11
	R-L-P-135F	Rural	53,23	556	Reseal at 5% surface crack	Preventive	1,80	80,91
	I-L-G-72D	Interurban	123,49	1525	Reseal at 5% surface crack	Preventive	5,56	86,46
	R-L-P-177E	Rural	64,18	385	Cold Mill and Inlay 70mm	Structural	6,07	98,04

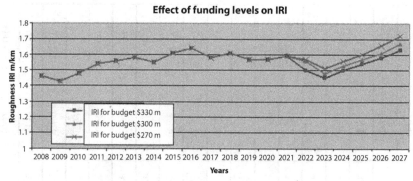

Figure 19.2 Effect of funding levels on IRI. After [23].

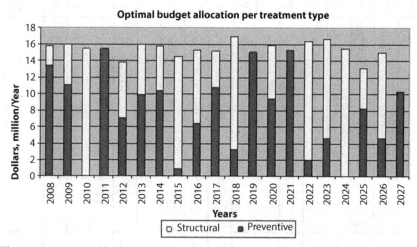

Figure 19.3 Optimal budget allocation per treatment type for combined program. After [23]

References for Part Three

1. Haas, R., W.R. Hudson, and J.P. Zaniewski, *Modern Pavement Management*, Krieger Publishing, Florida, 1994.
2. Haas, Ralph, "Life Cycle Management of Road Assets," Notes Prepared for Master class/workshops, Sydney, Melbourne and Brisbane, October 2011.
3. Monismith, Carl, "Evolution of Long-Lasting Pavement Design Methodology: A Perspective," Distinguished Lecture, *Proceedings*, Intl Symposium on Design and Construction of Long-Lasting Asphalt Pavements, Auburn University, June 2004.
4. AASHTO, "Mechanistic – Empirical Pavement Design Guide (MEPDG)," NCHRP Project 37-A, Washington, D.C., 2008.
5. Haas, Ralph, S. Tighe, G. Doré and D. Hein, "Mechanistic-Empirical Pavement Design: Evolution and Future Challenges," *Proceedings*, Transp. Assoc. of Canada Annual Conf., Saskatoon, October 2007.
6. Asphalt Pavement Alliance (APA), "Perpetual Asphalt Pavements: A Synthesis," Lanham, Maryland, 2010.
7. Cheetham, Alan, "Auto-Adaptive Pavement Performance Prediction Methodology," *Proceedings*, Vol. 2, pp 552-567, Fourth Int. Conf. on Managing Pavements, Durban, May 1998.
8. Applied Research Associates, Inc. (ARA), "Guide for Mechanistic-Empirical Design of New and Rehabilitated Pavement Structures," National Cooperative Highway Research Program (NCHRP) Project 1-37A, Transportation Research Board, National Research Council. National Academies, Washington, D.C., 2004.
9. AASHTO, "Mechanistic-Empirical Pavement Design Guide," A Manual of Practice, Washington, D.C., 2008.
10. Schwartz C.W., R. LI, S.H. Kim, H. Ceylan, K. Gopalakrishnan, "Sensitivity Evaluation of MEPDG Performance Prediction," National Cooperative Highway Research Program (NCHRP) Report 1-47, Transportation Research Board, National Research Council, National Academies, Washington, D.C., 2011.
11. Juhasz, Marta, Mark Popik and Susanne Chan, "Sensitivity of Pavement ME Design™ to Climate and Other Factors," *Proceedings*, Canadian Technical Asphalt Association, pp 239-253, St. Johns's, November 2013.
12. Ministry of Transportation Ontario, "Pavement Design and Rehabilitation Manual," SDO-90-01, Toronto, Ontario, 1990
13. Haas, Ralph, "The ICMPA7 Investment Analysis and Communication Challenge for Road Assets," Prepared for the 7th Int. Conf. on Managing Pavement Assets, Calgary, June 2008 (see www.icmpa2008.com).
14. Hall, Kate, J. M. Conner, M. I. Darter and S. H. Carpenter, "Development of an Export System for Concrete Pavement Evaluation and Rehabilitation," *Proceedings*, Second North American Conf. on Managing Pavements, Toronto, November 1987.

15. Transportation Association of Canada, "Pavement Asset Design and Management Guide," Ottawa, December 2013.
16. Haas, R., M.A. Karan, A. Cheetham, and S. Khalil, "Pavement Rehabilitation Programming: a Range of Options," *Proceedings*, First North American Pavement Management Conference, Toronto, Canada, March 1985.
17. Chan, W.T, T.F. Fwa, and C.Y. Tan, "Road Maintenance Planning Using Genetic Algorithms: Formulation," ASCE, J. Tramp. Eng., 120: 693-709, 1994.
18. Pilson, Charles, W.R. Hudson, and V. Anderson, "Multiobjective Optimization in Pavement Management by Using Genetic Algorithms and Efficient Surfaces," Transportation Research Board, TRR. 1655/1999, Washington, D.C., 2009.
19. Mathew, Binu Sara and Kuncharia P. Isaac, "Optimization of Maintenance Strategy for Rural Road Network Using Generic Algorithms," Intl. Journal of Pavement Engineering, Vol. 15, Issue 4, 2014.
20. Adeli, Hojjat, "Neural Networks in Civil Engineering: 1989-2000," Computer Aided Civil and Infrastructure Engineering, Blackwell Publishers, 16 126-142, 2001.
21. Kikukawa, S. and R. Haas, "Priority Programming for Network Level Pavement Management," *Proceedings*, Workshop on Paving In Cold Areas, Tsukuba, Japan, October 1984.
22. Smeaton, W.K., M.A. Karan, and R. Haas, "Determining the Most Cost-Effective Combination of Pavement Managements and Rehabilitation for Road and Street Networks," *Proceedings*, First North American Pavement Management Conf., Toronto, Canada, March 1985.
23. Chamorro, Alondra, Ehab Kamarah, Mohab El-Hakim and Wail Menesi, "ICMPA7 Challenge," Unpublished Report, Department of Civil and Environmental Engineering., University of Waterloo, 2008.
24. Haas, Ralph and Afrooz Aryanm "Cross Optimization," Final Report to Ministry of Transportation Ontario, Canada, July 31, 2013.

Part Four

STRUCTURAL DESIGN AND ECONOMIC ANALYSIS: PROJECT LEVEL

20

A Framework for Pavement Design

20.1 Introduction

Pavement design has evolved from an empirical and experience base to methods which attempt to combine mechanistic analysis with empirical aspects. Essentially though the concepts articulated in earlier works [1] and subsequent books [2,3] remain valid and define the overall pavement design process which is the first step in project level PMS.

The pavement design phase of a pavement management system in that earlier work had three major components:

1. Input information,
2. Generation of alternative design strategies, and
3. Analysis, economic evaluation, and optimization.

The framework for these components is intended to be applicable to both flexible and rigid pavements.

In the following chapters we pay attention to the MEPDG, for reasons subsequently explained, but precede this with a summary of basic structural response models, characterization of design inputs, and variability, reliability and risk and generating alternative design strategies in the design phase of pavement management.

20.2   Focus on the MEPDG

From 1996 to the time of this book, almost all national and state resources for pavement design research in the USA have been used in the development of a mechanistic-empirical pavement design method [4], called the Mechanistic-Empirical Pavement Design Guide (MEPDG). Since there is still no perfect mechanistic design method, the need exists for empirical calibration. Mechanistic models can only predict behavior or pavement response which, when carried to a limit, results in distress such as fracture, distortion, or disintegration. This step of relating primary behavior to damage is always empirical. Fatigue, for example, as a function of stress or strain is not determined by a natural law but is historically estimated by Miner's hypothesis [5].

Moving from distress to performance, or what the MEPDG terms "functional performance," requires another empirical step to combine and relate all distress to pavement roughness, then to ride/user evaluation in the form of present serviceability index (PSI) [6].

By way of background, the NCHRP project for the MEPDG was first awarded to an excellent research team headed by Drs. Gary Hicks, Carl Monismith, and Frank McCullough. Unfortunately, they resigned early in the project. A second contract was awarded to another prominent team headed by Drs. Mike Darter and Matthew Witczak. The project write-up at that time recognized the empirical necessity for adjusting calibration and transfer functions to deal with largely non-mechanistic parts of the problem. The research has been on-going since 1998 and, while an MEPDG framework was assembled and published in 2004 [4], changes, improvements, and upgrades are still ongoing at the time of this book. It is estimated, by the authors, that more than $30 million have been expended to date by state DOTs, FHWA, AASHTO, and NCHRP in the effort.

While the MEPDG is not really a project level PMS, it does have by inference significant impact on pavement management in DOTs. Rather than describe individual parts of the MEPDG, Chapter 21 summarizes some of the many pages of related material in the Guide reviewed by the authors. Project level pavement design as a part of PMS is more or less dormant in many highway agencies in favor of historical AASHTO Guide methods or the more complex MEPDG procedures. Significant attention in this book to the MEPDG should in no way detract from various pavement design methods used in Europe, Africa, South America, Asia, Australia, New Zealand, or other countries. An indication of the range of methods currently in use is found in the recently published *Pavement Asset Design and Management Guide* [7], which includes a scan of design methods used by Canadian provinces and territories. Of the twelve agencies responsible for pavement networks, eight are using the AASHTO 93 method, two are using the Shell Method (which is based upon elastic layer analysis) [8], and two are using either standard pavement sections or the State of Alaska Design Method. Unfortunately, the scope of this book has precluded giving deserved attention to these methods, but readers may consult such references as [9,10], published in the United States of America; [11], published in India; [12], published in Australia; and [7,13], published in Canada.

20.3   Basic Structural Response Models

Basic structural response models that are used in pavement design around the world include elastic layer analysis (such as in Chev5L [14], BISAR [8], or ELSYM5 [15]), thin plate theory [16], and numerical methods such as in ILLI-PAVE [17]. These methods are summarized in [3]. MEPDG introduces models of layered elastic analysis but not a new theory, and the pavement community has been working to develop software packages that incorporate this approach: one such program is JULEA [18].

Three other new models or formulations are described in the following:

ISLAB 2000 is a finite element formulation [19], which solves Westergaard's thin plate theory. While ISLAB adds useful details like discontinuities in the solution, it is based on Westergaard. PCC pavement design in MEPDG uses this formulation.

HIPERPAV, High Performance Paving Software, according to the developer Transtec Inc, HIPERPAV [20], is user-friendly, Windows-based software designed to assess the influence of pavement design, concrete mix design, construction methods, and environmental conditions on the early-age behavior of Portland cement concrete pavements. Planners, designers, contractors, and suppliers can use HIPERPAV, which is supported by FHWA. During the planning stage, it can be used to develop quality control specifications. Pavement designers can use it to optimize the design variables and guarantee long-term performance while maximizing economy. Contractors can use HIPERPAV to predict potential damage and determine how to prevent it; and with HIPERPAV, suppliers manage the temperature of the concrete based on their mix designs and specific climate and project conditions.

EICM, the enhanced environmental climate model, is a combination of theory and empirical equations that attempts to predict the complex variation of moisture content and temperature in a pavement structure. According to [21], the EICM is a one-dimensional coupled heat and moisture flow model initially developed for the FHWA and adapted for use in the MEPDG. It was developed under NCHRP Projects 1-37A and 1-40. In the MEPDG, the EICM is used to predict or simulate the changes in behavior and characteristics of pavement and unbound materials in conjunction with natural cycles of environmental conditions that occur over many years of service.

A summary listing of more well-known computer based packages for mechanistic analysis is provided in [22], as shown in Table 20.1. Note that this listing also identifies international contributions.

20.4   Characterization of Design Inputs

A basic background for establishing the inputs to project level pavement design are presented in [3]. These inputs include the following:

- Material properties, which are largely established by empirical methods but fundamental methods are increasingly coming into practice. There is an evolution in measurement procedures and some aspects are included in the MEPDG.
- Traffic inputs have traditionally been characterized in terms of Equivalent Single Axle Loads (ESALs). The MEPDG uses load spectra as traffic input. However, because of the complexity involved, many agencies will likely continue to estimate load spectra by simulation rather than direct measurement.
- Environmental inputs, which include temperature variations, moisture variations, rainfall, solar radiation, freeze and thaw, and contaminants.

Table 20.1 Summary of some computer-based analytical solutions for asphalt concrete pavements. After [22]

Program and Ref.	Theoretical Basis	No. Layers (max)	No. of Loads (max)	Program Source	Remarks
CHEV5L Internal Rpt	MLE	5	1	Chevron Research	Can not calculate subgrade strain
BISAR[1] [12, 22]	MLE	5	10	Shell International	The program BISTRO was a fore-runner of this program
ELSYM[2] [13, 22]	MLE	5	10	FHWA (UCB)	Widely used MLE analysis program
PDMAP (PSAD) [14, 22]	MLE	5	2	NCHRP Project 1-10	Includes provisions for iteration to reflect non-linear response in untreated aggregate layers
JULEA[3]	MLE	5	4+	USACE WES	Used in Program LEDFAA
CIRCLY[4] [15, 22]	MLE	5+	100	MINCAD, Australia	Includes provisions for horizontal loads and frictionless as well as full-friction interfaces
VESYS [16,22]	MLE or MLVE	5	2	FHWA	Can be operated using elastic or viscoelastic materials response
VEROAD [17, 22]	MLVE	15 (resulting in half-space)		Delft Technical University	Viscoelastic response in shear; elas-tic response for volume change

(Continued)

Table 20.1 Summary of some computer-based analytical solutions for asphalt concrete pavements. After [22] (Continued)

Program and Ref.	Theoretical Basis	No. Layers (max)	No. of Loads (max)	Program Source	Remarks
ILLIPAVE [18,22]	FE		1	University of Illinois	
FENLAP [19,22]	FE		1	University of Nottingham	Specifically developed to accommodate non-linear resilient materials properties
SAPSI-M [20,22]	Layered, damped elastic medium	N layers resting on elastic half-space or rigid base	Multiple	Michigan State Univ./Univ. of California Berkeley	Complex response method of transient analysis-continuum solution in horizontal direction and finite element solution in vertical direction

MLE – multilayer elastic; MLVE – multilayer viscoelastic; FE – finite element

1. Current version is described in "Shell Pavement Design Manual" (Personal Computer Version SPDM-PC), by J.N. Preston 1996, Shell, Delft
2. ELSYM5 is available from McTRANS™ in Florida
3. JULEA is described in LEDFAA (Layered Elastic Design Federal Aviation Administration), User's Manual, FAA, Washington, D.C., 1995
4. CIRCLY4 is the current version, Wardle and Rodway, Proc., Transport 98, ARRB Transport Res., Victoria, Australia, 1998. Recently, Wardle, Rickards and Lancaster have adopted/modified CIRCLY4 in developing "HIPAVE" for the M-E design of heavy duty industrial pavements (sec ICAP10 Proc., Quebec, 2006)

- Interactions of actors always exist to greater or lesser degrees, but the state of pavement design still has a long way to go before these are adequately incorporated as inputs.

20.4.1   Materials Inputs

Many primary materials are used in pavements, as well as perhaps many more specialized materials. Among the primary materials are the following:

- Asphalt or bituminous binders and mixtures
- Portland cements and Portland cement concrete
- Granular base and subbase materials
- Cohesive and cohesionless soils
- Stabilizers such as lime, also in use as an anti-strip agent
- Reinforcement materials such as steel, various fibers, plastics, etc.
- Additives or modifiers such as reclaimed tire rubber, various polymers, etc.

A vast body of information exists on test methods for these materials and characterization of test results for use in design, construction, and maintenance. The major sources include the American Society for Testing and Materials (ASTM), with more than 13,000 Standards (www.astm.org); Reunion Internationale des Laboratoires et Exports des Materiaux, RILEM, with over 50 years in materials and structures tests and standards (http://rilem223dwh.isqweb.it); and others directly in the transportation field such as AASHTO.

A key aspect of any materials test and characterization is to simulate, if applicable, expected low and high temperatures, including rates of change, as well as low and high moisture contents and also rates of change. The intent is to simulate field conditions as closely as possible, recognizing that these can never be fully duplicated in a laboratory or even field test.

20.4.2   Traffic Load Inputs

Traffic loads for pavement design have traditionally been considered in terms of the number of equivalent single axle loads (ESALs) over a design period. The ESAL concept arose from the AASHO Road Test and is still widely used [23]. More recently, however, the concept of load spectra as inputs for pavement design, including individual wheel loads and tire pressures, has been developed for the MEPDG [24]. The fact that load spectra can only be measured on sites which have the specialized weigh-in-motion equipment, and the fact that few agencies have such equipment, is further discussed in Chapter 21.

Actual variation of axle loads on highways is substantial, from loaded trucks sometimes overweight in comparison to legal loads, to unloaded trucks. As well, speeds can vary substantially from free flow conditions to stop and go congestion. This is a major reason why traffic inputs to design can have a high variance and why ongoing evaluation of performance is necessary for any design updates and for rehabilitation interventions.

20.4.3   Environmental Inputs

Environmental conditions have a major effect on the characteristics of materials, pavement performance, and on construction and maintenance. In a direct sense, environmental conditions that usually have a major influence are temperature, moisture, and solar radiation. While materials characterization attempts to capture these conditions or effects as much as possible, as pointed out in Section 22.2, local or regional influences can add to the variation. Consequently, a common approach to performance prediction in design is to incorporate all these conditions into age as a surrogate.

20.4.4   Interactions

Interactions of input factors are always present, and while it would be desirable to incorporate these into pavement design, the challenge they present is major. For example, Figures 15.4 in Chapter 15 indicates that five major classes of factors, ranging from environment, to structure, to construction, to maintenance and traffic all have interactions that can affect performance. Moreover, several dozen sub factors within these factors have interactions. Consequently, most working design models incorporate an aggregation of factors, calibrated to local or regional conditions, and they use surrogates, such as age, as independent variables.

20.5   Variability, Reliability and Risk in Pavement Management

The ability to accurately predict pavement behavior and performance depends on the inherent variability of the many factors involved in design, construction, and in-service use. Sources of variation also include the amount and type of traffic and transfer functions that relate pavement response to traffic. Another major source of variation are the environmental factors and pavement performance since, in essence, there is variability in the factors that is not explained by the independent variables in the design equations.

20.5.1   Variance in Pavement Design

Uncertainty in the ability to predict pavement performance or service life is related to variability in materials characterization, traffic loads, and environmental factors and errors in the prediction models themselves.

The application of reliability concepts in pavement design was explored by researchers in the 1970s, and the 1986 AASHTO Guide for Design of Pavement Structures recognized the effect of several variance terms in design predictions. It defined reliability as the probability that pavement sections will withstand the actual number of axle loadings and environmental conditions that will be applied over the design life of the pavement.

20.5.2   Formulation of Pavement Reliability

The application of a reliability formulation or concept to pavement design involves selection of a pavement structure for which the number of loads that can be carried through its service life, N_t, is equal to the number of loads that are actually carried, n_t. Otherwise the pavement is overdesigned or under designed.

Of course the reality is more complex in the sense that there are other predicted and unobserved variables versus what actually occurs. These individual variables can be accumulated into an overall variance, S^2_0. Thus the reliability for a design is related to standard (normal curve) deviation of the design process, z_R as:

$$Log\ (F_R) = -z_R\ S_0$$

where F_R is the reliability term.

Various reliability levels for z_R and process standard deviation are listed in the AASHTO Guides. For example, the 1993 AASHTO Guide [23] suggests the levels for various functional classes of highways as shown in Table 20.2.

The inherent variability in pavement materials, environmental factors, construction, maintenance, traffic, and other factors makes it impossible to perfect the performance of any pavement with 100 percent certainty. Including the reliability concept in design provides a more rational

Table 20.2 Suggested levels of reliability for pavement design. After [23].

Functional Classification	Percent Reliability (%)	
	Urban	Rural
Interstate and Other Freeways	85–99	80–99
Principal Arterials	80–99	75–95
Collectors	80–95	75–95
Local	50–80	50–80

approach to the expectation of performance and in turn to a more reliable economic analysis of the design alternatives available.

20.5.3   Reliability Concept in the MEPDG

In 1987, Irick stated, "The AASHO Road Test is virtually the only performance study that has produced comprehensive field data on performance prediction variance" [25]. That statement was essentially still applicable to the 1993 AASHTO Guide. Now, the MEPDG as a successor incorporates a reliability calculation process, but its reliability, in turn, remains to be tested.

The basic formulation in the MEPDG is that design reliability is based on the probability that each of the key deterioration measures of distress and roughness will be less than the selected critical level over the design period. Each measure or factor is characterized by a mean and variance.

A level of reliability for each factor is chosen by the designer, commonly according to suggested levels in the MEPDG for different classes of highway. These levels can be the same for all factors or they can vary. For example, a terminal level for IRI of 2.5 m/km may be associated with a chosen reliability level of 90%. But a reflective cracking amount for rehabilitation of say 100% may only be accompanied by a reliability level of 50%.

It is important that the application and verification of reliability applications from the MEPDG be evaluated in the next decade.

It should be noted that some of the commercial systems outlined in Part Six incorporate the concepts of risk and reliability.

20.6   Generating Alternative Design Strategies

Selection of a structural thickness is an essential component of pavement design. However, a more comprehensive concept of a design strategy would include alternative pavement types such as flexible, rigid and composite, layer materials, expected construction and maintenance policies or procedures, rehabilitation or preservation treatment alternatives, and on-going performance evaluation procedures.

Starting with all the design inputs, objectives, and constraints, a process for generating alternative pavement design strategies is described in [3] for new designs, rehabilitation alternatives, prevention/preservation treatments, reconstruction alternatives, and evaluation process alternatives. This is the process followed for several decades in various design procedures including early ones such as [26] and later such as [27].

20.6.1   Generating Structural Design and Overlay Alternatives Example

The Ontario Pavement Analysis of Costs (OPAC) system, termed "OPAC 2000," is based on a mechanistic-empirical analysis for flexible pavements and on the 1993 AASHTO Guide for rigid pavements [27]. It includes a comprehensive reliability analysis and life cycle economic analysis module. The framework for generating structural and future overlay strategy alternatives is a good example of the procedure noted in Chapter 17. A schematic summary is shown in Figure 20.1

20.6.2   Materials Alternatives

The example in Figure 20.1 is concerned with generating structural layer alternatives. However, there may also be materials type alternatives within the structural alternatives. For example, the base materials in a flexible pavement may be unbound granular, asphalt-treated, or cement-treated. Furthermore, the asphalt for the treated base may be an emulsified material or asphalt cement.

The foregoing suggests that available material types need to be combined with incremental layer thickness alternatives to generate a number of possible design strategies in any given situation.

20.6.3   Construction and Maintenance Policy Alternatives

A comprehensive design strategy should include the expected construction and maintenance policies. For example, the expected construction policies might include variations or limits in layer thicknesses, as-built roughness, traffic-handling method, and schedules. Any deviations from these expectations, such as different sources of materials or increased costs, should be recorded in the data base.

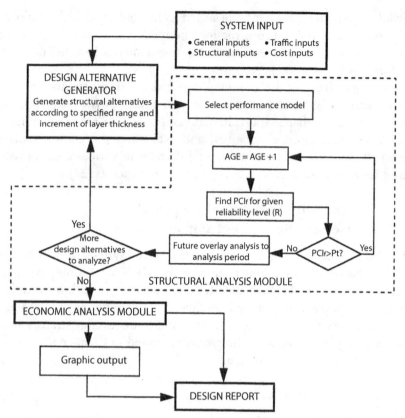

Figure 20.1 Framework for generating structural design alternatives in OPAC 2000. After [27]

Expected maintenance policies in generating alternative design strategies might include the levels or frequency of maintenance treatments. These policies vary with the type of facility, traffic volumes, budgets, and even with accounting procedures for capital and maintenance expenditures.

20.6.4   Pavement Evaluation

The expected evaluation of pavements during their service lives should be a consideration at the design stage. Information on performance or deterioration allows for updates in design predictions and for developing better performance models.

The expected evaluation policies of types of measurements, frequencies, calibrations, data transfer, etc. should also be communicated to the agency's design office or team responsible. In this way, any design changes can be made and, in turn, communicated back to the evaluation office or team. This can be particularly important in larger agencies where these are separate design and other offices. For example, evaluation or monitoring may be in a materials office.

20.6.5   Alternative Designs in the MEPDG

The MEPDG process does not involve the generation of alternative design strategies like the approach noted in the foregoing sections. That approach consists of generating and evaluating all of the alternatives, applying any constraints such as minimum initial service life, and carrying out life cycle economic analysis to identify one or more of the most economically attractive strategies. The intent is to identify an optimal or near optimal strategy.

In the MEPDG the approach is to carry out analysis of one or more trial designs for a specific set of inputs and designated criteria. These criteria, for flexible pavements, would usually include maximum roughness in terms of terminal International Roughness Index, maximum distress in terms of bottom up and top down fatigue cracking, thermal cracking, and rutting. If a trial design fails a criterion, design inputs are usually changed. For example, the Pavement Asset Design and Management Guide [7] provides an example where the first trial design did not meet the required permanent deformation/rutting and top down fatigue cracking. So the asphalt grade was changed and the trial design came closer to meeting the criteria.

The MEPDG example in [7] used 25 truck classification and other input parameters, plus 32 drainage and materials inputs. While over 300 inputs can be used, this may be a typical design scenario. However, even though various trial designs can be analyzed and evaluated for their life cycle costs, whether the design selected is optimal or even near optimal is not indicated.

21

The MEPDG Process for Pavement Design

21.1 Introduction

In the past decade comprehensive details have been introduced into pavement design in the Mechanistic-Empirical Pavement Design Guide (MEPDG) [4,23]. The MEPDG involves many but not all components of design covered in [3]. In particular the MEPDG integrates the design inputs, structural response models, structural analysis for asphalt concrete and Portland cement concrete pavements, and to some degree rehabilitation.

The analysis and design of flexible and rigid pavements has been well covered in [3,10,28]. Because the MEPDG has been in development and calibration since the late 1990s and has consumed large amounts of resources, and because the method is comprehensive and deals with all pavements in similar ways, we have chosen to treat the MEPDG as a whole in this chapter. In that way it supplements historical references [3,28].

MEPDG is a project-level design method, not a project-level PMS, and after completing the MEPDG process the user only obtains an entry point to PMS. The selected "design" must first be constructed, then the design and as-constructed data must be entered into the PMS database for use over the ensuing life cycle. An overview of the MEPDG design process is provided in Figure 21.1 [AASHTO 04]. It incorporates three stages: evaluation, analysis, and strategy selection. The latter term is arguable in view of the discussion in Section 20.6.

The flexible part of MEPDG is more analysis than design and can involve over 350 variables. This is an enormous increase in the number

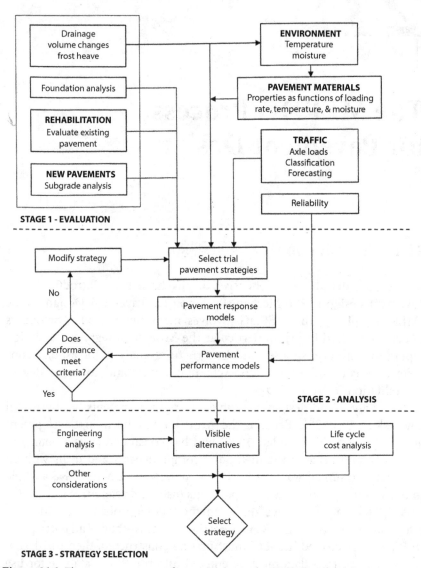

Figure 21.1 Three-stage pavement design process in the MEPDG. After [4]

of variables normally considered in pavement design. Structural analysis for Portland cement concrete pavement involves over 65 variables. Many claims are made for the validity of MEPDG, yet at the time of this book, proof remains to be established that use of MEPDG produces better pavement performance than previous methods. Whereas the initial attempts by NCHRP were to produce a mechanistic pavement design method, it became evident early (2000-2001) that a purely mechanistic design method was impossible because mechanistic models do not predict pavement performance, only pavement behavior. This has been shown continually over the last half century and has been documented in the work of Yoder and Witczak [28], Carey and Irick at the AASHO Road Test [6], and clearly in the AASHO Road Test results [29], as well as early work in pavement management by [1,2,30,31] and many others including even earlier work by Westergaard [32].

21.2   Calibration Issues

It is difficult to understand why an NCHRP project charged with improving pavement design in 1998 would undertake a multi- million-dollar mechanistic project, particularly when the entire approach for pavement management was originally funded by NCHRP in 1968-1972 [1]. This action was taken in response to the very fact that it was impossible to mechanistically design a pavement to fit all conditions. There is no doubt that good mechanistic equations were developed in MEPDG [4]. But the Guide to date has been calibrated mostly with field data that was taken from the long-term pavement performance monitoring project [33,34]. Early LTPP documents made it clear that the data expected from nationwide field studies were never going to be accurate on a project-by-project basis. The intended approach was to develop a fractional factorial experiment design which would collect data from up to 1,000 pavement sections in the field and would use a factorial statistical analysis [35] to define the influence trends of the basic factors while minimizing the effect of errors in any specific measurement.

It is clear from discussions over the years with members of the MEPDG project team that section-by-section calibration field data was filtered so that "smoothed" curves were used for the calibration rather than often erratic raw data. In addition, the sections selected for calibration were the very best available and poorer sections were not acceptable. One dramatic change introduced in the MEPDG is the use of load spectrum for the calculation and accumulation of stresses under various individual loads. Such data is only available in the LTPP dataset on a limited number of pooled fund sites with high precision bending plates or load cells.

Even after national calibration for the MEPDG, it is stated in [4] that user agencies must calibrate the methodology to local conditions. As of this writing, some 30 states have undertaken calibration projects. Because it is patently impossible to obtain all 300 plus variables for individual existing pavement sections calibration efforts to date seem to have considered limited sets of what are considered important or significant variables. The danger in this is missing significant interactions. The only reasonable explanation is that traditional engineering training can tend to ignore the statistical variability of inputs to detailed mathematical equations.

For a clearer understanding of this idea, look at the numbers involved. A proper calibrated or sensitivity analysis for the flexible pavement guide would require looking at each variable on at least three levels, low, medium, and high. The number involved would then be 3 to the power 350, an impossibly large number. To bring the idea further into perspective, taking 100 variables at only three levels gives 3 to the power 100, which equals 5.15×47 zeros. Taking only 10 variables at three levels yields 3 to the 10th power, which equals 59,049 combinations. Even this small number of 10 variables at three levels requires an impossibly large number of combinations (calculations) in a calibration or sensitivity analysis.

As a result, agencies attempt to calibrate MEPDG over only a few selected variables such as thickness, strength, and/or Equivalent Single Axle Loads (ESALs). These are the same variables used for pavement design since the AASHO Road Test in the early 1960s. Historically, most states have found it difficult even to produce good estimates of ESALs for pavement design. One can imagine how difficult it will be to produce realistic axle loads spectra involving dozens of variables and axle counts over time in 6–10 individual axle categories on each in-service pavement section.

21.3   MEPDG Software

The version of software available at the time of this writing is called "Pavement ME." It superseded DARWin-ME in 2013.

An undated (approximately 2010) article from AASHTO entitled DARWin-ME [36] describes the "next generation of pavement design software which builds on the NCHRP Mechanistic-Empirical Pavement Design Guide and expands and improves the features," with the following statement:

For many pavement engineers it is a paradigm shift away from a nomograph-based design to one based on engineering principles and mechanics. Instead of entering basic site and project information into an equation and getting an empirically based pavement design output, the engineer can use detailed traffic, materials, and environmental information to assess the short and long term performance of a pavement design using nationally and/or locally calibrated models.

This quote appears to contradict the status of pavement design as it has existed for the last five decades. All previous AASHTO methods were based on engineering principles and mechanics. The rigid pavement methods in particular have always been based on Westergaard's theory of slab behavior [32] modified by measurements of strain, precise load measurements, and field measurements of soil subgrade and subbase characteristics at the AASHO Road Test [37]. It is not easy to understand why such statements would emanate from an organization like AASHTO, especially in view of the excellent research results funded by states through AASHO at the Maryland Road Test [38], WAASHO [39], and AASHO Road Tests [29].

In reality, the MEPDG builds on original mechanistic design principles but adds significant numbers of details, which may well challenge agencies with limited budgets.

21.4   Levels of Use in the MEPDG

The MEPDG literature outlines possible use at three levels.

Level 1. Detailed level – Where most of the input variables are obtained by measurement.

Level 2. Intermediate level – Where many of the variables are default values or dummy variables.

Level 3. Practical/default level – Where most of the inputs are default values and only the basic variables actually are measured.

Based on the authors' experience, the default level (Level 3) will be predominant. Experience with the initial AASHO Design Guides show that even after many years the default values for structural number of the asphalt surface, the crushed stone base, and the gravel subbase, 0.44, 0.14, and 0.11 respectively, were still being used by many state and international agencies, even though clear guidance was given in the Guides on how to obtain measured values.

21.5   Good Design is Not Enough - Life Cycle Pavement Management is Also Needed

As pointed out in cited project level PMS literature, it has been clear since the late 1960s that good design is not enough, no matter how many variables are inputs because variables are only future estimates of what is expected to occur in the pavement including traffic. Design thicknesses and material properties must be realized in construction. Since all of these factors are truly "variable," we must construct the pavement to the best of our ability using a reasonable method, and then we must "manage" that pavement for the rest of its source life as outlined in [3].

The MEPDG has little application at the network level of pavement management because the models and processes are too complex and time consuming for integration into network level pavement management. Agencies that use the MEPDG method for design need to connect to the network level of pavement management by entering the final design and as-constructed variables into the PMS database as they have in the past. There will not be room in the database for more than basic variables such as traffic, thickness, materials strength, etc. The means and variability of all other variables can be entered in a trailer file.

Suffice it to say the MEPDG does not lend itself to pavement management at the project level and even less at the network level. It does offer future potential when computational power and measurement techniques become more practical. The MEPDG documentation constitutes hundreds of pages and is continuously updated. We will therefore cover only the main points and suggest that the reader reference the latest version of the Guide for study.

21.6   Summary of the MEPDG for Flexible Pavements

21.6.1   Basic Mechanistic Principles

While the MEPDG employs new and complex models, the basic model is elastic layered theory developed by Burmister in 1945 [40] and used by Yoder and Witzcak among many others [28]. A number of computer programs are available to do elastic layered analysis: the one widely used in MEPDG is known as JULEA [18,41]. The JULEA program uses typical input data such as elastic modulus for each of the unbound layers, dynamic modulus for the hot mix layer, layer thicknesses, Poisson's Ratio, tire pressure, and contact area of the tire applying the load. Elastic layered theory in general contains eight basic assumptions which are required but not always realistic in the pavement world. They are covered well by Yoder and Witczak [28].

21.6.2   Design Inputs in MEPDG

The MEPDG uses over 350 design inputs which reportedly represents the highest precision available in input data and in calculation of design results. The input data includes laboratory materials tests and many field variables such as 20 to 30 traffic variables. The example listed in Section 20.6.5 used 25 traffic and 32 materials inputs. Default values constituted the remaining variables. As MEPDG unfolded, it was soon recognized that most pavement design agencies and consultants were not likely to obtain 350 inputs for design. They therefore designated three levels of data input: Level 1, high precision; Level 2, some estimates and default values; and Level 3, mostly default values. Basically the same calculations are used for all three levels. Only the accuracy of inputs is different.

21.6.3   Traffic Inputs for MEPDG

NCHRP 1-37A [4] outlines the following traffic data needed as MEPDG inputs:

- Traffic Volume (base year information)
- Two-way annual daily truck traffic (AADTT)
- Number of lanes in the design direction
- Percent of trucks in the design direction
- Percent of trucks in the design lane
- Vehicle operational speed
- Traffic volume adjustment factors
- Monthly adjustment
- Vehicle class distribution
- Hourly truck distribution
- Traffic growth factors
- Axle load distribution factors
- Number of axles per truck
- Axle configuration
- Wheel base

Traffic data requires the use of traffic load spectrum in lieu of equivalent single axle loads (ESALs). ESALs were developed soon after the AASHO

Road Test and are the most widely used measure of heavy traffic damage in the world. While state-of-the-art traffic counting and weigh-in-motion equipment can provide the spectrum data for a few sections, such equipment is not widely enough distributed in states and provinces to provide the spectrum data on all pavements being designed. In fact, a check of most existing pavement management systems in U.S. states and Canadian provinces will show that age is a more commonly used measure than even ESALs because of lack of data.

Most pavements managers are familiar with ESALs and have a "feel" for them. Traffic spectra are more rigorous but this involves so many numbers for various axle loads that the accuracy of predictions in the future can be questionable. Pavement management systems themselves generally combine raw field data into manageable indexes such as condition index, roughness index, traffic ESALs, etc.

As you can imagine, the equations needed to combine all of these variables to predict the traffic spectrum for MEPDG are long and complicated and will not be reproduced here.

21.6.4   Climate Inputs

Another major change in MEPDG from previous methods is climate. Climate is incorporated into the pavement design analysis through the Enhanced Integrated Climatic Model (EICM). This model was developed by the Federal Highway Administration, the University of Illinois, the United States Cold Regions Research and Engineering Laboratory, and Texas A&M University [42]. The model is merged into MEPDG through a computer program which summarizes data that permeate many calculations including resilient modulus and other material characteristics presumably affected by changes in moisture within the materials over time and depth. It also evaluates the effect of freezing and thawing and soil moisture conditions and analyzes how resilient modulus changes the computation of the pavement response. The latitude, longitude, elevation, and depth of water table in feet are required inputs for selecting the "station" within the MEPDG to generate the climatic file required for each design.

21.6.5   Pavement Performance

The MEPDG concept of pavement performance considers functional performance, structural performance, and safety. According to the MEPDG, the most important concerns are functional and structural performance. Structural performance includes the analysis of fatigue cracking and permanent deformation for flexible pavements. Functional performance is related to the service the highway provides to the user. The most important functionality condition is serviceability as indicated by ride comfort or quality. The original serviceability concept was expressed in terms of the Present Serviceability Index (PSI) [6]. International Roughness Index (IRI) is the measure selected by MEPDG. Equation 1.20 in NCHRP 1-37A gives the general model for roughness. ·

21.6.6 Problems Observed in Implementing MEPDG in State DOTs

There are many implementation studies by states trying to use the MEPDG. A typical example and a good reference for using the MEPDG was prepared by Mallela, von Quintus, *et al.* and the Ohio DOT [43]. A typical statement from the report follows:

> Based on a laboratory evaluation it was determined that larger aggregates combined with aged materials tend to have high modulus values at high temperatures. However, neither the E* Bar nor the frequency sweep at constant height (FSCH) could correctly rank the permanent deformation characteristics of six HMA mixtures tested. Both tests were sensitive to the permanent deformation characteristic for the mixture evaluated. This study points to the deficiency of using E*Bar with a vertical test for rutting characterization.

This is just one such statement which points to the difficulty of using complicated tests in a realistic way.

22

The MEPDG for Design
of New and Reconstructed
Rigid Pavements

22.1 Introduction

The MEPDG combines the design procedures for new and reconstructed jointed concrete pavements (JCP) and continuously reinforced concrete pavements (CRCP) into an iterative approach. The performance measures considered in this method include joint faulting and transverse cracking for JPCP and punch-outs for CRCP and International Roughness Index for both pavement types. The designs that meet the applicable performance criteria at the selected reliability level are then considered feasible from a structural and functional standpoint and can be further evaluated for factors such as life-cycle cost analysis and environmental impacts.

While there are no fundamental differences in the way pavements are designed, new or reconstructed, a practical aspect for reconstruction is the potential reuse or recycling of materials from existing pavements structures.

22.2   Overview of the Design Process

The overall iterative design processes for JPCP and CRCP are illustrated in Figures 22.1 and 22.2 respectively. The MEPDG for rigid pavements can consider many structural layer arrangements and design features, including

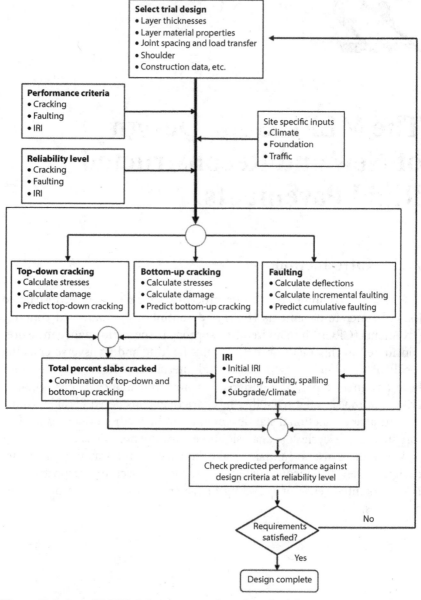

Figure 22.1 Overall MEPDG design process for JPCP. After [4]

joint spacing, dowels, tied PCC shoulders, widened slabs, base-type, and drainage. The process is done iteratively to identify a design that satisfies the performance criteria (i.e., joint faulting, slab cracking, punch-outs, and IRI) over the analysis period. A trial design includes all details needed for evaluation using the procedures in the Guide for pavement layers, joint

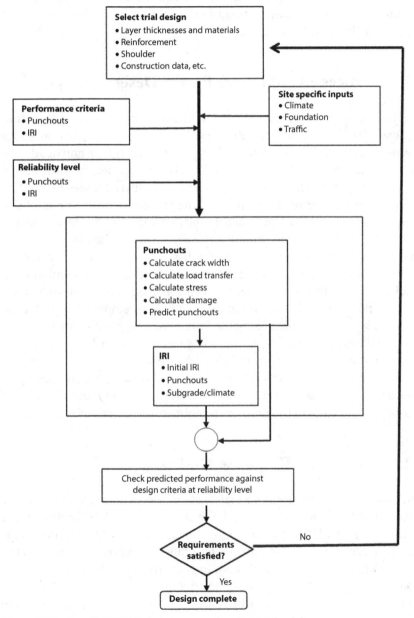

Figure 22.2 Overall MEPDG design process for CRCP. After [4]

design, reinforcement design, and material properties. The designer must also provide inputs for project site conditions, including subgrade properties, traffic, and climatic data. Other variables include initial IRI, estimated month of PCC paving, estimated month pavement open to traffic, and estimated permanent curl/warp of the PCC slab. As for the flexible design procedure, three levels of data inputs are considered. The highest level 1 includes measurement and testing of material properties. The lowest level 3 uses mostly default values.

22.3   Processing of Inputs for the Design Analysis

The raw design inputs are processed to obtain monthly values of the traffic, material, and climatic inputs needed in the design evaluation which consist of average hourly number of single, tandem, tridem, and quad axles in each axle weight category, in 13 axle weight categories, for each month of the analysis period; temperatures at 11 evenly spaced nodes in the PCC layer for every hour of the available climatic data for a minimum of one year's weather data; average monthly relative humidity for each calendar month; PCC strength and modulus at each month of the analysis period; monthly average moduli values of the base layer; and monthly average effective subgrade modulus of reaction based on subgrade resilient moduli. Traffic calculations are the same as previously described for flexible pavements. The monthly layer moduli and the hourly temperature profiles in the PCC layer are obtained using the Enhanced Integrated Climatic Model (EICM) which is part of the Guide software [42]. Major layer types to be considered in the procedure are PCC slab, asphalt stabilized base, cement stabilized base, other chemical layers, and unbound aggregate base/subbase and subgrade soil. Each of these materials are inputs and the effect of seasonally changing temperature and moisture conditions are calculated.

22.4   Structural Response Models

The pavement response such as stresses and deflections under the influence of traffic and load are calculated using the ISLAB2000 finite element structural model [19,44] to compute critical stresses and deflections rapidly. However, since the incremental monthly design procedure used in the MEPDG requires hundreds of thousands of stress and deflection calculations to compute monthly damage for the different loads, load positions, and equivalent temperature differences over many years, the requisite computations would take literally days to complete using finite element methods. To reduce this computer time to a practical level, neural networks (basically regression equations) were developed based on the finite element results to compute critical stress and deflections quickly. The trial designs are analyzed by dividing the analysis period into monthly segments using load spectra the same as for flexible pavements. Within each month for each increment all other factors that affect pavement response damage are held constant including PCC strength and modulus, base modulus, subgrade modulus, joint load transfer both transverse and longitudinal, and finally base erosion and loss of support for CRCP.

Thus within each increment a critical stress or deflection is calculated as well as the damage incurred in that time increment. Damage is summed over all increments as an output at the end of each month by the Guide software. Calibrated distress prediction models were developed using LTPP data and other long-term pavement performance data obtained for a wide range of JPCP and CRCP pavements located in a variety of climatic conditions and subject to various traffic and environmental loads.

For details beyond the scope of this book, the reader should refer to the literature outlined in the references or the latest timely version available.

23

Rehabilitation of Existing Pavements

23.1   Introduction

In addition to preservation/preventive maintenance treatments, the life cycle of a pavement may include one or more rehabilitation treatments. The AASHTO Guides broadly classify rehabilitation into overlays, non-overlay methods, and reconstruction [23,45]. Many agencies still use the AASHTO Guides; however, numerous other methods are in use as discussed in Chapter 20. Design guides and resources are available from:

- Transportation Association of Canada
- AASHTO
- AUSTROADS
- NCHRP
- American Concrete Pavement Association
- Portland Cement Association

According to NCHRP 1-37A [4], the MEPDG covers mechanistic-empirical design procedures for hot mix asphalt (HMA) overlays of flexible, semi-rigid, composite, and rigid pavements. The HMA overlay design process covers:

- ˙ HMA overlay of existing HMA surfaced pavements, both flexible and semi-rigid.
- HMA overlay of existing PCC pavement that has received fractured slab treatments; crack and seat, break and seat, and rubblization.
- HMA overlay of existing intact PCC pavement (JPCP and CRCP), including composite pavements or second overlays of original PCC pavements. Note that there is no specific overlay design procedure for JRCP. However, some recommendations are provided for approximate overlay design of JRCP considering reflection cracking and distress in the HMS overlay.

23.2   MEPDG Suggested Evaluation Data for Pavement Rehabilitation

MEPDG discusses 11 categories of data needed for evaluation of pavements for rehabilitation.

1. Traffic loads.
2. Pavement condition (e.g., distress, smoothness, surface friction, and deflections).
3. Condition of pavement-shoulder interface.
4. Pavement design features (e.g., layer thicknesses, structural characteristics, and construction requirements).
5. Material and soil properties.
6. Traffic volumes and loadings.
7. Climatic conditions.
8. Drainage conditions.
9. Geometric factors (e.g., bridge clearance).
10. Safety aspects (e.g., rate and location of accidents).
11. Miscellaneous factors (e.g., utilities and clearances).

The Guide also suggests evaluating the following major aspects of the existing pavement.

- Structural adequacy (features that define response to traffic loads).
- Functional adequacy (surface and subsurface properties that define the smoothness or frictional resistance of the pavement surface).

- Subsurface drainage adequacy.
- Material durability.
- Shoulder condition.
- Variation of pavement condition or performance within a project.
- Miscellaneous constraints (e.g., bridge and lateral clearance and traffic control restrictions).

23.3   MEPDG Rehabilitation Design with HMA

Figure 23.1 outlines the HMA rehabilitation design process. Structural design of feasible rehabilitation strategies is Step 6 of the procedures shown. Note that the last three important steps are shown as optional in the method. Why this is so for a highly detailed method is not clear.

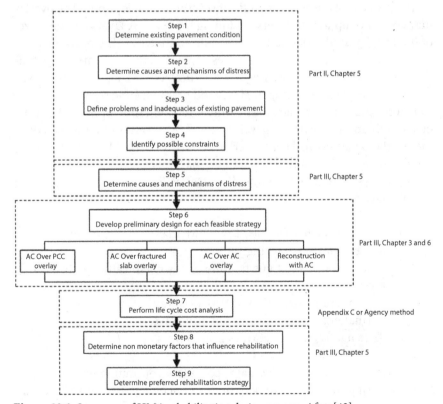

Figure 23.1 Summary of HMA rehabilitation design process. After [42]

The MEPDG says that HMA overlay is a candidate rehabilitation strategy for either HMA or PCC surfaces, including the following:

- Conventional flexible pavements – thin HMA layer over granular base and subbase.
- Deep strength HMA pavements – thick HMA layers over granular base and subbase.
- Full-depth HMA pavements – flexible pavement consisting only of HMA layers.
- Semi-rigid pavements – HMA surfaced sections having some type of chemically stabilized layer.
- Composite pavements – HMA surface over PCC. These may include previous HMA overlays of original PCC pavements.
- PCC pavements – Jointed plan concrete pavements (JPCP), jointed reinforced concrete pavements (JRCP), and continuously reinforced concrete pavements (CRCP).

The procedures provide analysis of several overlay options. The overlay may consist of up to four layers, including three HMA layers and one layer of unbound granular or chemically stabilized material. The procedure can also assess the effects of various types of pre-overlay treatments such as cold milling of existing HMA layers, fracture/rubbilizing of existing PCC layers, and in-place recycling of HMA and granular base layers.

Figure 23.2 is a flow chart for pavement rehabilitation options. The various combinations of existing pavements and pre-overlay treatments reduce the overlay analyses to HMA overlay of HMA surfaced pavement, fractured PCC Pavement, and/or intact PCC pavement.

The analysis predicts the same distresses as for new and reconstructed flexible pavements:

- Load associated fatigue of the HMA layers, both top-down and bottom-up cracking.
- Load associated fatigue fracture of any chemically stabilized layer.
- Permanent deformation in HMA layers.
- Permanent deformation in unbound layers.
- Thermal fracture in HMA surface layers.

The HMA over PCC analysis also considers continuing damage of the PCC slab using the rigid pavement performance models. The analyses can also address reflection cracking of joints and cracks in PCC pavements

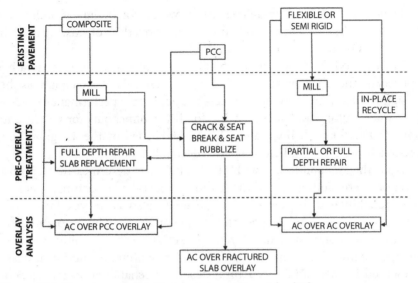

Figure 23.2 Flow chart of rehabilitation design options. After [42]

and thermal and load cracking in HMA surfaced pavements. However, it should be noted here that the reflective cracking models incorporated in the Guide were based strictly on empirical observations, not rigorous M-E analysis. Finally, the predicted distresses are used to estimate International Roughness Index (IRI), and estimate functional performance to consider along with the specific distresses.

23.4   MEPDG Rehabilitation Design with PCC

The MEPDG also contains procedures for rehabilitation of existing flexible, rigid, and composite pavements with Portland cement concrete (PCC). Lane additions and widening of narrow lanes can also be considered.

The MEPDG covers procedures for several PCC rehabilitation strategies:

1. Design of concrete pavement restoration (CRP) for JPCP.
2. Design of unbonded JPCP or CRCP overlays over existing rigid and composite pavements.
3. Design of bonded PCC overlays over existing JPCP or CRCP.
4. Design of conventional JPCP or CRCP overlays over existing flexible pavements.

In addition, general guidelines are provided for design of additional traffic lanes. The design of ultra-thin concrete overlays of existing asphalt pavements is not covered in the Guide.

The method for designing rehabilitated pavements requires an iterative approach. The designer selects a trial rehabilitation and then analyzes the design to determine whether it meets the applicable performance criteria (i.e., joint faulting and slab cracking for JPCP, punchouts for CRCP, and smoothness for both JPCP and CRCP) established by the designer. This process is repeated until an acceptable design is found.

Note that rehabilitation with JPCP or CRCP describes the topmost layer of the rehabilitated pavement and not the type of existing pavement to be rehabilitated. The design procedures can also use recycled materials if the recycled material properties can be characterized by the parameters used in the design and the recycled material meets durability requirements. Figure 23.3 summarizes the PCC rehabilitation design process presented in this Guide. Structural design of rehabilitation is Step 6 of the figure.

Figure 23.4 presents a summary of rehabilitation design. Figures 23.3 and 23.4 have common factors for rehabilitation. There are, however, important differences in rehabilitation design strategies, which is the term used in the figures. Again, it is arguable whether these are truly strategies in the sense described in Chapter 20, Section 20.6.

23.5   Concrete Pavement Restoration (CPR) of JPCP

Several non-overlay rehabilitation treatments can be used on existing JPCP to restore functionality and structural capacity. In the MEPDG, a package of rehabilitation treatments (CPR) is considered to restore a deteriorated JPCP to adequate functionality and to restore load carrying capacity.

Some commonly used CPR treatments are presented in Table 23.1. The performance of the individual CPR treatments listed is directly related to proper identification of condition and of treatments needed to prevent further deterioration, timing of the CPR work, and quality of construction and materials.

The Guide says that properly designed and constructed CPR may reduce pavement deterioration and prolong pavement life but the information presented is empirical, not M-E. It also references external guidelines on CPR.

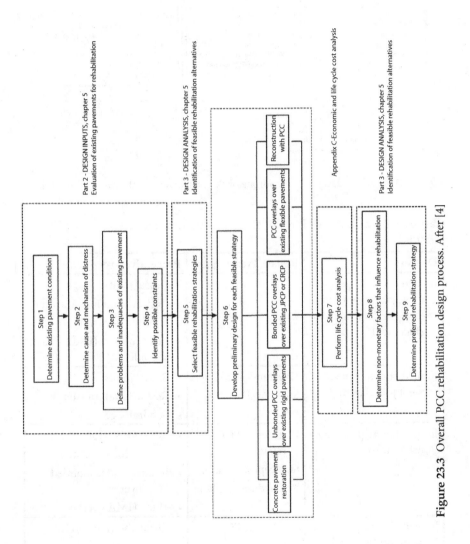

Figure 23.3 Overall PCC rehabilitation design process. After [4]

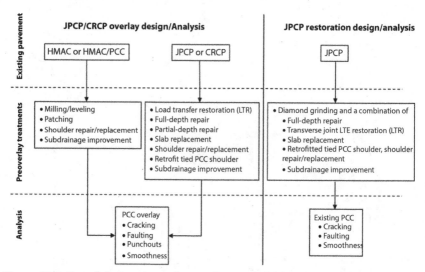

Figure 23.4 Overall design process for major PCC rehabilitation strategies. After [42]

Table 23.1 Candidate CPR repair and preventive treatments for existing JPCP. After [4]

Distress	Repair Treatments	Preventive Treatments
Jointed concrete pavement pumping (and low joint load transfer efficiency)	—	• Reseal joints • Restore joint load transfer • Subdrainage • Edge support (tied PCC shoulder)
Jointed concrete pavement joint faulting	Diamond grinding Structural overlay	• Reseal joints • Restore load transfer • Subdrainage
Joint concrete pavement slab cracking	Full-depth PCC repair Slab replacement Replace/recycle lane	• Retro fit PCC shoulder • Restore load transfer • Bonded and unbonded PCC overlays • Thick HMA overlays
Jointed concrete pavement joint or crack spalling	Full-depth PC repair Partial-depth repair	• Clean and reseal joints
PCC disintegration (e.g., D-cracking and alkali-silica reaction [ASR])	Full-depth repair	• Thick hot mix AC overlay • Unbonded PCC overlay

23.6   Models, Algorithms, and Transfer Functions of the MEPDG

There are dozens of equations, variously termed models, algorithms, and transfer functions in the Guide. Most of these models occupy a full page in the Guide [24] report and contain numerous calibration coefficients, adjustment factors, and other variations which must be determined from a multitude of input variables and submodels. These models and transfer functions are incorporated in a computer program, but getting the necessary inputs to it is a huge undertaking. In general, in the MEPDG, the term "mechanistic" describes a framework that requires multiple models, which must all be validated and calibrated, and large data entry converted into indices before it can be used effectively by any highway agency. Even with easily useable software for such a complicated system, a typical user will be challenged to understand the basics of the system and generate the input data required in a cost effective way. It is almost certain that default values will be widely used except for the basic variables always used in design, such as thickness, strength, and subgrade quality. It is impossible to validate or calibrate or determine the sensitivity of models containing more than 300 input variables. The second, third, and fourth order interactions of real world data can invalidate any one or two factor evaluation.

23.7   Quality of Calibration Data and Factor Adjustments

One of the major issues with establishing a valid MEPDG is the quality of data and information available to do the empirical adjustments needed for the mechanistic equations. Two figures taken from the MEPDG project [4] illustrate the problem.

Figure 23.5 shows the relationship for predicted versus average estimated measured rutting. Note that neither of these values are real values: one is predicted and the other is estimated based on other data. Even the relationship between the predicted and the estimated based on modeling are relatively poor. A large percentage of the data range in the bottom third of the graph and much of the data is at least 50% off the correlation average. Figure 23.6 plots top-down cracking versus fatigue damage at the surface of the HMAC layer. Most of the data is in the bottom third of the plot of longitudinal cracking but the curve misses most of the available data. It is easy to understand the difficulty in calibrating equations using data with such variability.

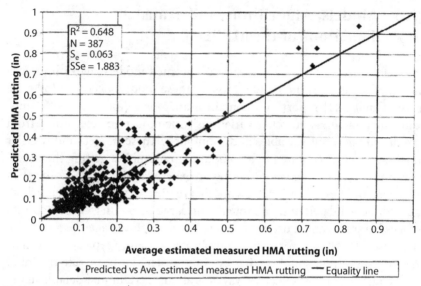

Figure 23.5 Nationally calibrated predicted versus estimated measured asphalt rutting. After [4]

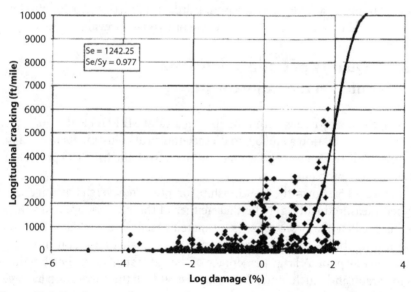

Figure 23.6 Top-down cracking versus fatigue damage at surface of HMA layer. After [4]

23.8   AASHTO Manual of Practices

AASHTO [24] is a well-organized description of the details needed to perform mechanistic-empirical design. Although it does not say so in the manual itself, the same team that prepared [4] apparently prepared it.

The Manual contains information about almost every conceivable pavement factor except for the following 13 factors as noted on page 23, which are not considered in this method.

1. Friction or skid resistance and noise
2. Single and super-single tires
3. Durability and mixture disintegration
4. Volume change in problem soils
5. Asphalt treated permeable bases
6. Geogrids and other reinforcing materials
7. Semi-rigid pavements
8. Pavement preservation programs
9. Staged construction
10. Ultra – thin PCC overlays
11. JRCP jointed reinforced concrete pavements
12. Early aged Portland cement concrete pavement opening to traffic
13. Inter-face friction of HMA overlay and existing PCC pavement

It also references over 60 AASHTO or other specifications and Guide documents required to get input data for the methods. This Manual of Practice, in its latest version, is what state DOTs will likely use to implement the Guide.

One of the major problems shown on page 42 of the Manual (Figure 23.7) plots the comparison of measured and predicted longitudinal cracking (top-down).

The figure indicates an R^2 value of 0.544 but the only plot on the figure is the equality line which is not a best fit line. More than 95% of the data points fall below measured and/or predicted cracking of 3300. By looking at the two axes, most of the points fall directly on or very close to one of the axes. About 40% of the points show near 0 predicted cracking, while measured cracking ranges up to 2800 feet per mile. Also about 30% of the data points show predicted cracking up to 1800 feet per mile when measured cracking is 0. These findings do not add confidence to the use of this relationship. If the top 8 or 10 points were eliminated, the R^2 would be far lower than 0.54.

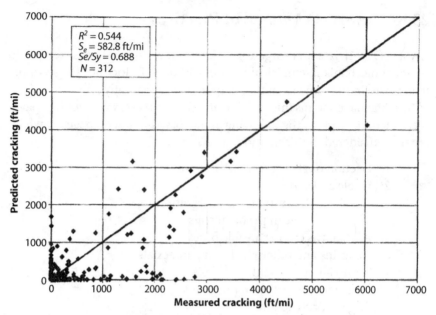

Figure 23.7 Comparison of measured and predicted lengths of longitudinal cracking (top-down cracking) resulting from global calibration process. After [23]

24

MEPDG in Practice

24.1   Use of the Guide in Pavement Management

The good news is that without a doubt the MEPDG references are the best summary of design related factors ever assembled for all pavement types and should form the basis for a more practical design procedure with better calibrated relationships in the next 10 to 15 years. Even then the complexity and the cost of obtaining needed inputs may continue to be prohibitive in time and money for most agencies including state DOTs.

In any event the Guide has limited direct application in pavement management. However, if an agency uses the method, then design results should be entered into the project level PM database file identified with that section and possibly in a trailer file for all the design variables because there is not sufficient capacity in most existing data files for 350 variables. These entries in the project level file can be used for future comparison and correlation. Over several years this will build a dataset that can be used for more valid future calibration as the number of sections and the related dataset accumulate.

24.2   MEPDG Offers a Roadmap to Improvement

Thirty factors for improvement are outlined below by the MEPDG flexible pavement team headed by Dr. Matthew Witczak and the rigid pavement team headed by Dr. Mike Darter [4]. Both are highly qualified and rank among the very best in the field. Their teams included 32 well-qualified engineers on the flexible team and 26 on the rigid team. They did an admirable job in putting together an overall framework and structure for an empirically modified mechanistically based design method. However, these 30 factors, as subsequently listed, clearly point out that the method is not a practical, fully functioning design method for use in pavement management systems.

We sincerely applaud the team on their work to fulfill the objectives of the NCHRP project. However, we feel strongly that a better approach to the improvement of pavement performance, which was the ultimate goal of the project, is to implement a strong PMS (pavement management system). No matter how much good data is manipulated with models, reliable prediction from that data for 20 to 50 years in the future remains impossible.

24.3   MEPDG Research Team's Perspective on Guide Improvements

The NCHRP report 1-37A is one of the earliest, most complete summaries for MEPDG. As a setting for the discussions to follow, summarized from the Forward of that report, are the development needs as outlined by the research team itself.

Some of these issues have been addressed in subsequent activities but many have not. Because of the exceptionally large volume (over 3,000 pages) of information related to the Guide and various subsystems, it is difficult to put your finger on the exact situation at any given time.

The thirty factors for improvement, as noted previously and summarized from [4], are:

1. The MEPDG assumes that proper inputs are used often through default values (not a very mechanistic approach).
2. Any model or algorithm can be replaced; however, changes to models for distress and smoothness may require recalibration.
3. It is recommended that results from NCHRP 9-23 (nearing completion) incorporate modifications to the EICM.

4. An enhanced database may lead to better calibrated and more accurate climate models. Design reliability procedures should be considered as a place holder for more comprehensive procedures.

5. Calibrations-validation is needed for prediction models for level 1, 2, and 3 inputs. It was only possible to demonstrate so far that detailed inputs seem to produce more accurate modeling results.

6. Several major sensitivity studies are still needed for various models.

7. Improved accuracy is needed in the LTPP database for validating distress/smoothness models.

8. It is critical that trench studies be completed on certain LTPP flexible test sections.

9. Need to modify existing LTPP procedures to better identify longitudinal cracking.

10. Need to establish national center for the coordination of state calibration efforts.

11. Need to improve the accuracy of smoothness (IRI) models.

12. An enhanced validation effort for HMA pavements and overlays is greatly needed.

13. Trenching studies are needed on the LTPP sections for validation of rutting models.

14. Need to validate longitudinal surface (top down and bottom up) crack prediction models in the LTPP database.

15. Need to enhance/improve/update many existing models in MEPDG to improve accuracy.

16. The reflective crack model for HMA overlays is an empirical place-holder for future development.

17. The rutting model for HMA needs an empirical relationship to adjust the rutting as a function of depth.

18. Need to reduce the computational time for flexible pavement design.

19. Enhancements are needed to the Witczak *et al.* E* predictive model.

20. Initial calibration should be conducted on FEM (Finite Element Methods) technology for asphalt pavement systems.

21. The current design guide can only handle PCC overlay thicknesses of 6 inches and greater.

22. The method of incorporating permanent shrinkage into the permanent curl/warp model needs improvement.

23. Permanent curl/warp effective temperature difference needs further calibration and amplification.
24. The coefficient of thermal expansion/contraction (CTE) has never before been measured and used in design; therefore, much more information is needed.
25. The CRCP prediction of both crack spacing and crack width greatly need additional validation since they play a critical role.
26. More validation of crack deterioration model is needed.
27. An enhanced calibration/validation effort is greatly needed for rigid pavements.
28. There is a great need for additional PCC rehabilitated test section data.
29. Need to enhance/improve existing models to improve prediction accuracy.
30. Methods to estimate PCC mixture and construction factors for design are limited and rudimentary, thus needing significant improvement.

24.4   Practical Experience with MEPDG Flexible Pavement Models

To examine practical information about the value of the MEPDG, we corresponded with Mr. Harold Von Quintus in Austin, Texas [46]. He has worked extensively in all phases of flexible pavement design and rehabilitation for more than 30 years and was intimately involved in the original calibration of the models in MEPDG and in further calibration efforts in the DOTs of Arizona, Colorado, Wyoming, Missouri, Mississippi, Utah, and others. As a consultant he has also personally used the method. He said:

> In short, yes, I have used the method for multiple projects. It is like any other procedure. If the user does not understand the inputs, it will get misused. It typically focuses on the rehabilitation portion for most potential users because that is where most agencies are using the method. We often got significantly different results from the agencies' procedure. The question then is which is correct?

Mr. Von Quintus believes the MEPDG gives more accurate results than many local procedures because it provides better estimates of the in-place damage through deflection testing of the bound layers. He thinks this is a big advantage but "only if used properly." He pointed out that back calculation of elastic modulus is separate from the MEPDG and that only the resulting modulus values are used for the bound and unbound layers in design. Obviously the age-old problem of which back calculation method and what assumptions are used in the method remain. Mr. Von Quintus also pointed out that another big item that he focuses on for new design is how well the method reproduces actual observations. He says, "This is how I judge whether a procedure has value or is reasonably accurate." He has compared the standard errors of the methods with other existing methods and feels that the MEPDG is as good as, if not better than, those methods. But he also says, "Obviously this assumes the inputs have been correctly determined for both procedures." This is hardly a ringing confirmation of the value of the method since many millions of dollars and countless data values are required for using the method. Given the cost to date, it should in fact be much better than any previously existing method to justify the expenditure.

Von Quintus points out, as we previously have, that many of the values used in rehabilitation design are still default values embedded in the program. He points out that he does not use these values because his experience does not validate them. He also confirms that since MEPDG is elastic layer-based, all the assumptions that apply to elastic layer theory still apply to the MEPDG. It is also stiffness based where an elastic or dynamic modulus is a key parameter and that is the focus for inputs in level one use of the method. He says, "We know other important material properties not used in the method. For new design the program assumes that elastic modulus explains the difference in performance." It is his opinion that while stiffness is important, many other factors also contribute to pavement deterioration and there is a need to focus more on strength and permanent deformation factors. He points out that NCHRP project 9-30A focused on the use of permanent deformation parameters for HMA layers and clearly shows that these are more important than dynamic modulus in explaining rutting characteristics of HMA layers [21].

Mr. Von Quintus is an excellent pavement researcher and practicing engineer who obviously makes use of the MEPDG but points out that the basic needs for this method are the same for all other pavement design methods, which is to get good input data for use with the method. Clearly it is much more difficult to get good data for 300 plus variables than it is for 10 or 15 variables.

24.5 Use of MEPDG for Rehabilitation and Overlay Design

Others who have used the MEPDG for rehabilitation design find that it has some weak links. For example, the reflection cracking regression equation has been found lacking and will be replaced in the future. There is an insufficient link between the cracks in the existing surface of the old pavement and how they reflect to the surface of an overlay. This is clearly the same problem that we have always had with overlay design. More importantly, there is no tie between the coefficients of the reflection cracking regression equation and different reflection cracking mitigation techniques. NCHRP is giving consideration to trying to improve this mechanistic-empirical-based reflection cracking model in the future. There are also some types of cracks that MEPDG does not predict. For example, the transverse cracks typically seen in hot dry climates such as Arizona, New Mexico, and West Texas which are probably more related to shrinkage than to cold temperatures are not predicted at all by the MEPDG. In addition, the MEPDG program can calculate percentage cracking values greater than 100% which of course is impossible. Efforts are being made to correct these problems.

24.6   Mechanistic-Empirical Pavement Design Software

The current (as of 2013) software for the MEPDG is Pavement-ME. It is the successor to DARWin-ME both of which build on the MEPDG and expand and improve the features [36]. According to Scofield, DARWin-ME is a production ready software tool which supports day-to-day operations of public and private pavement engineers. The software is mainly aimed as a tool for AASHTO state DOTs. A detailed examination of the software is beyond the scope of this book. Readers will have to evaluate the software for themselves.

According to presentations made at various AASHTO meetings, the software supports the MEPDG Guide published in 2008. It purportedly covers 17 pavement design situations including new concrete and asphalt pavements and various types of asphalt and concrete overlays. It is designed to operate on Microsoft Window OS and contains over 20 engineering modules. It does not purport to mimic the MEPDG but says it provides state-of-the-art software "consistent" with AASHTO's MEPDG. There will undoubtedly be updates to this software in the future and readers should avail themselves of the latest version for study.

24.7   Summary

We would like to point out that these same types of problems have always existed with prior methods and creating a method with 300 plus input variables does not seem to have solved the problem. If the user does not understand the MEPDG inputs, the likelihood is that it will get misused. The same can be said for any other existing pavement design equation and method. The problem is that trying to understand 300 plus variables is extremely more difficult than understanding 10 or 12 variables.

Studies by ARA Inc., *et al.* find the method useful, but they are probably among the top 1% of people in the world knowledgeable about the method [21,43,47]. They say that it is difficult to use the method if people do not understand it and the inputs. Only time will tell whether or not this extremely complicated method will be of practical use to routine designers. The authors own preference is that simpler methods be used for initial design and then the pavement be managed with a good pavement management system over time; regardless of the number of input variables or how accurately they are predicted, they will change in the next 10 to 50 years and therefore the results will be different than originally predicted.

All of these factors suggest that use of the complex MEPDG will not replace pavement management; rather the need for pavement management must be highlighted where design using MEPDG or any other method only serves as the initial first step or starting point of life cycle management.

25

Economic Evaluation of Alternative Pavement Design Strategies and Selection of an Optimal Strategy

25.1 Introduction

Many references present a solid and detailed background for economic evaluation, including cost and benefit factors, methods of evaluation, and examples of the project level [3]. At the network level most software has used approximate incremental benefits (INCBEN) or marginal cost-effectiveness analyses for maximizing benefits and/or optimizing costs [3,48]. In addition to the network priority programming examples in Part Three, several references were found related to bridges [49,50]. Saad used INCBEN to rank improvement alternatives in decreasing order of their incremental benefit-cost ratios for bridges. He states that INCBEN internally adds do-nothing alternatives to bridges without considering their consequences. This method of benefit cost analysis should be examined carefully before use.

Saad has developed what he terms a multi-media bridge management system called "Manager." He does not compare the benefits of INCBEN versus true optimization but merely states that he chose to use INCBEN. AgileAssets software discussed in Part Six uses true optimization, not

INCBEN, and does so with multi-constraints and over a multi-year horizon [51]. The best information available on Deighton software indicates they use INCBEN.

AgileAssets and possibly others also incorporate an efficiency frontier analysis into their optimization as part of their pavement management software [52]. Other references to efficiency frontier relate primarily to its use in optimizing stock portfolios [53].

According to Investopedia the efficiency frontier concept was introduced by Harry Markowitz in 1952 and is a cornerstone of modern portfolio theory. It is also applicable to pavement management or other resource allocation problems such as physical asset management. In physical asset management, risk relates to the portion of the budget allocated to a particular project, and reward is the improvement and benefits gained in the overall network.

25.2   Consideration of Environmental Costs in Selecting Alternative Strategies

Environmentally friendly materials and decisions may be more expensive initially than standard practice. Also it is unknown whether such materials will perform as well as standard, less environmentally friendly materials. A large amount of information has been gathered about the use of Recycled Asphalt Pavement (RAP) and its reuse in new asphalt surfaces. Similarly, there is much information on "Green Roads," which can be found on Google or Wikipedia. Environmental costs are often discussed with concern but little data or experience with alternative materials is available. The good news is that traditional life-cycle cost models (LCCA) can readily handle such materials if and when reliable life cycle environmental costs and performance data becomes available. The cost will ultimately be determined by contractor bid prices, but that usually only occurs after the cost analysis to select the optimum options takes place. The other good news is that any deviations will show up in the resulting network level PMS performance data over time. The best way to obtain this longer term cost and performance data is through a pavement management system. In fact, it may be the only practical way.

25.3   Weighing Costs versus Environmental Benefits

While cost and performance will sort themselves out over time, analysis and field data cannot define the value that agencies and the road using public place on the associated environmental benefits. That must be done based on perceived public support or, in some cases, where support is strong enough that it may result in changes in state or federal law mandating a specific level of change or the use of environmentally friendly materials such as RAP. This change already manifests itself by some agency mandates that a fixed percentage of RAP be incorporated in asphalt concrete pavements, overlays, and/or new surfaces. The resulting level of performance will be defined by field observations and distress measurements after five or six years.

This information will help to define the performance of environmentally friendly materials but does not define the environmental benefits. Little published data is found to define such benefits. So far, agencies are using anecdotal estimates or scores, such as in "Green Roads," of the perceived value of reducing air pollution or conserving natural resources such as aggregate. After research is done to define the global benefits of reducing things like air and noise pollution or neighborhood disruption, these benefits can be appropriately weighted based on public opinion or legislative mandate and used in the economic analysis.

25.4   Unique and/or Unpredictable Cost Factors

Unpredictability of costs is a concept which needs to be considered in future pavement management systems. Not all cost factors are predictable. Examples of unpredictable costs are inflation, interest rates, and fuel or asphalt prices. For many years interest rates were in the 4–6% range, but since 2008 in the United States, the Federal Reserve Bank has artificially depressed interest rates by selling large amounts of bonds and lending to public banks at near 1% interest. This type of policy changes the effect of interest or cost of money on life-cycle cost analysis.

Likewise, crude oil prices have suffered violent swings over the past 20 years. For many years they remained constant or increased gradually year by year. But in recent years, global conflicts and new production methods like fracturing shale have caused large gyrations in oil prices. This affects asphalt and fuel prices cyclically from about $3 to $4 per gallon. Since both asphalt and fuel are major costs in highway procurements, construction, and maintenance, these price fluctuations impact pavement management. Such unpredictable cost variations can be handled by annual evaluation and reanalysis using adaptable network level PMS software.

25.5   User Costs

User costs are important in economic evaluation of pavements. Unfortunately many highway agencies still do not accept the validity of user costs and thus seldom use them in decision making. But there are exceptions such as the New Jersey DOT [54]. Highway agencies have been beset by shrinking budgets and refusal by law makers in the United States to increase tax revenue for highways. This financial stress creates more concern for agency costs themselves and less concern with user and environmental costs. Decision makers are hard to convince that all highway costs are ultimately paid by the public users through taxes and/or by self-funded user costs such as travel delay, air pollution, and noise. A specific example can be cited in TxDOT. When highway funds are plentiful, TxDOT usually maintains older pavements with 1½ to 2 inch overlays on a cycle of seven to ten years or when excessive cracking is manifest. When funds shrink, however, TxDOT typically shifts to using single surface seals with coarse cover stone instead of thin overlays. This results in doubling tire noise pollution for vehicle occupants and neighbors and creates a large increase in broken windshields due to flying stone.

These and other needs for future improvement in PMS practices are discussed in Chapter 45. Readers and pavement managers may well supplement this writing with new research results as they emerge over the next 10 to 15 years in line with future research outlined in Chapter 45.

25.6   Selection of an Optimal Strategy

Primarily project level optimal strategy selection is addressed in detail in [3]. The optimal project strategy is the one with minimum uniform annual cost or discounted present worth of costs over the performance life of the pavement, including periodic maintenance/overlay and rehabilitation actions; or the maximum performance in terms of cost-effectiveness obtainable for available funds. These are adequately dealt with by [3].

What has become more important over the last 20 years is the optimum annual allocation of scarce financial resources for building and maintaining a large pavement network over a 5, 10, or 20 year horizon to meet agency objectives. These objectives are usually to provide maximum overall performance on the network for minimum costs or for available funds. As pointed out previously, funds available to DOTs for the last 8 to 10 years have been less than the minimum needed to provide good performance. Agencies use several ways to allocate their available resources: worst first, prioritization, incremental benefit (INCBEN), and true optimization. Historically, many agencies tabulate the current condition of their pavement network divided into pavement sections, then list these sections in order with the pavement section in the worst condition at the top of the list. They then allocate funds to make necessary repairs or rehabilitation to these bad pavements as far down the list as funds will allow. This is not the best use of funds.

Other agencies and various commercial PMS software prioritize sections based on a variety of criteria including severity of condition, traffic, loads to be carried, highway classification, etc. While certainly better than "worst first" funding, such prioritization is still far from optimal use of funds. Unfortunately some agencies and some software providers call prioritization "optimization" and some cannot tell the difference. Optimization whether by INCBEN methods or true optimization are far superior to the first two approaches but they are also more difficult and absorb more time. The best method is of course true optimization [55].

25.7   Summary

If pavement managers wish to develop their own optimization method, they will need to study the current methods at the time of their need. Users who want to purchase or lease commercially available PMS software should query the potential software providers and ask for a specific demonstration and validation of the optimization methodology used in the software. Some providers have historically used prioritization but called it optimization. Most, at least, currently use INCBEN for optimization. The best software providers can also optimize over a 10 or more year horizon with multiple constraints [51]. Part Six herein provides details, examples, and contacts.

References for Part Four

1. Hudson, W.R., F.N. Finn, B.F. McCullough, K. Nair, and B.A. Vallerga, "Systems Approach to Pavement Design, Systems Formulation, Performance Definition and Materials Characterization," Final Report, NCHRP Project 1-10, Materials Research and Development Inc., March 1968.
2. Haas, Ralph and W.R. Hudson, *Pavement Management Systems*, McGrawHill, New York, 1978.
3. Haas, R, W. R. Hudson, and J.P. Zaniewski, *Modern Pavement Management*, Krieger Publishing, 1994.
4. "Guide for Mechanistic-Empirical Design of New and Rehabilitated Pavement Structures," NCHRP Project 1-37A, Final Report, Part 2, Design Inputs, Chapter 4: Traffic, Transportation Research Board, Washington, D.C. March 2004.
5. Lalanne, Christian, *Fatigue Damage*, ISTE, Wiley Publishers, 2009.
6. Carey, W.N. and P.E. Irick, "The Pavement Serviceability-Performance Concept," HRB Bulletin 250, Highway Research Board, 1960.
7. Transportation Association of Canada, *Pavement Asset Design and Management Guide*, TAC, Ottawa, Canada, 2013.
8. Shell International Petroleum Company Ltd., "Shell Pavement Design Manual: Asphalt Pavements and Overlays for Road Traffic," London, U.K., 1978.
9. Papagiannakis, A.T., and E.A. Masad, *Pavement Design and Materials*, Wiley, 2008.
10. Huang, Yang H., *Pavement Analysis and Design*, Second Edition, Prentice Hall, 2003.
11. Kumar, R. Srinivasa, *Pavement Design*, Gopal Books University Press, 2013.
12. Jameson, Geoff, "Technical Basis of Austroads Guide to Pavement Technology - Part 2: Pavement Design", ARRB, Melbourne, Australia, 2013.
13. Transportation Association of Canada, *Pavement Design and Management Guide* Transportation Association of Canada, Ottawa, Canada, 1997.
14. Chevron Oil Company, Development of the Elastic layer Theory Program, 1963.
15. Ahlborn, G., "Elastic Layer System with Normal Loads," Institute for Transportation and Traffic Engineering, University of California, Berkeley, 1972.
16. Sargious, M. *Pavements and Surfacing for Highway and Airports*, John Wiley and Sons, New York, 1975
17. Gomez-Achecar, M. and M.R. Thompson, "ILLI-PAVE-Based Response Algorithms For Full-Depth Asphalt Concrete Flexible Pavements," Transportation Research Board, URL: http://www.trb.org/Publications/Pages/262.aspx, 1987.
18. Uzan, J. "Jacob Uzan's Linear Elastic Analysis-JULEA," Technion-Haifa, Israel, 2001.
19. Applied Research Associates, "ISLAB2000 Finite Element Modeling Software," http://www.ara.com/products/ISLAB2000.htm, ISLAB Sales, 2000.

20. The Transtec Group, Inc, HIPERPAV III, High Performance Paving Software, http://www.hiperpav.com/index.php?q=node/1, with the FHWA, 1996, 2011.

21. NCHRP, "Calibration of Rutting Models for Structural and Mix Design," NCHRP Report 719, Project Number: 09-30A, Transportation Research Board, Washington, D.C., 2012.

22. Haas, Ralph, Susan Tighe, Guy Dore and David Hein, "Mechanistic - Empirical Pavement Design: Evolution and Future Challenges", Proc, Transp. Assoc. of Canada Annual Conf., Saskatoon, Sept., 2007

23. AASHTO, *Guide for Design of Pavement Structures*, Washington D.C., 1993.

24. American Association of State Highway and Transportation Officials, "Mechanistic-Empirical Pavement Design Guide, A Manual of Practice," Interim Edition, July 2008.

25. Irick, P., W. R. Hudson and B. F. McCullough, "Application of Reliability Concepts to Pavement Design," Proc., Sixth Int. Conf. on the Structural Design of Asphalt Concrete pavements, Univ. of Michigan, 1987.

26. Hudson, W.R. and B. F. McCullough, "Flexible Pavement Design and Management Systems Formulation," NCHRP Report 139, 1973

27. He, Zhiwai, Gerhard Kennepohl, Ralph Haas and Yinyin Cai, "OPAC 2000: A New Pavement Design System," Proc., Transp. Assoc. of Canada Annual Conf., Charlottetown, Oct., 1996.

28. Yoder, E.J. and M.W. Witczak. *Principles of Pavement Design*. 2nd Ed., Wiley & Sons, Inc. New York, 1975.

29. Highway Research Board, "AASHO Road Test Principal Relationships – Performance Versus Stress, Rigid Pavements," Highway Research Board Special Report 73, 1962 (W.R. Hudson and F.H. Scrivner).

30. Wilkins, E.B, "Outline of a Proposed Management System for the CGRA Pavement Design and Evaluation Committee," *Proceedings*, Canadian Good Roads Association, 1968.

31. Hutchinson, B.G. and R.C.G. Haas, "A Systems Analysis of the Highway Pavement Design Process," Highway Research Record No. 239, Highway Research Board, 1968.

32. Westergaard, H.M., "Theory of Concrete Pavement Design," *Proceedings*, Highway Research Board, 1927.

33. Darter, M.I., "Long-Term Pavement Performance Program Highlights: Accomplishments and Benefits 1989-2009, Contributions to Pavement Design and Management," Report no. FHWA-HRT-10-071, August 2010.

34. Mallela, J., *et al.*, "Implementation of the AASHTO Mechanistic-Empirical Pavement Design Guide for Colorado," Final Report No. CDOT-2013-4, CDOT, July 2013.

35. Anderson, V.L. and R.A. McLean, *Design of Experiments, a Realistic Approach, Volume 5,* Marcel Dekker, Inc., New York, 1974.

36. DARWin-ME, Mechanistic-Empirical Pavement Design Software, contact V Scofield, AASHTO Project Manager, vscofield@aashto.org

37. Highway Research Board. "The AASHO Road Test: Report 5-Pavement Research," HRB Special Report 61-E, Highway Research Board, 1962.

38. Highway Research Board, "Maryland Road Test One MD: Final report, Effect of Controlled Axle Loadings on Concrete Pavement," Highway Research Board Special Report 4, 1952.

39. Highway Research Board, "The WASHO Road Test, Part 2: Test Data analysis and Findings," Highway Research Board Special Report 22, 1955.

40. Burmister, D.M., "The General Theory of Stresses and Displacement in Layered Systems," Vol 16(2), Vol 16(3), Vol 16(5), Journal of Applied Physics, 1945.

41. Uzan, J., "Permanent Deformation in Flexible Pavements," Journal Transp. Eng., 130(1), 6–13, ASCE Technical Papers, 2004.

42. "Calibration and Validation of the Enhanced Integrated Climatic Model for Pavement Design," NCHRP Report 602, Project Number: 09–23, Transportation Research Board, Washington, D.C., 2008.

43. Mallela, J, et al., "Guidelines for Implementing NCHRP 1-37A M-E Design Procedures in Ohio: Volume 1 – Summary of Findings, Implementation Plan, and Next Steps," Applied Research Associates, Inc., in cooperation with the Ohio DOT and FHWA, State Job number 134300, Champaign, IL, 2009.

44. Transportation Research Board, "Development of Rapid solutions for Prediction of Critical CRCP Stresses," by Khazanovich, Lev, et al., TRR Vol. 1778, online date: January 2007.

45. AASHTO, *Guide for Design of Pavement Structures*, 4th Edition, with 1998 Supplement, Washington, D.C., 1998.

46. Personal communications with Harold Von Quintus in Austin Texas October 16, 2013.

47. Von Quintus, Harold L., J. S. Moulthrop, "Mechanistic-Empirical Pavement Design Guide Flexible Pavement Performance Prediction Models for Montana: Volume I Executive Research Summary," FHWA/MT-07-008/8158-1, August 2007.

48. McFarland, W.F. , J.B. Rollins, R.Dheri, *Documentation for Incremental Benefit Cost Techniques (INCBEN)*, FHWA/Texas Transportation Institute, 1983.

49. Farid, Foad, D.W. Johnston, B.S. Rihani, and C. Chen, "Feasibility of Incremental Benefit-Cost Analysis for Optimal Budget Allocation in Bridge Management Systems," Transportation Research Board, Issue Number 1442, 1994.

50. Saad, Ihab M.H., "Manager: Multimedia Bridge Management System," INCBEN, 47th ASC Annual International Conference, *Proceedings*, Associated Schools of Construction, 2011.

51. Galenko, Alexander, A. Bhargava, T. Scheinberg, "Asset Management Optimization Models: Model Size Reduction in the Context of Pavement Management System," 2013 IJPC – International Journal of Pavements Conference, São Paulo, Brazil, December 9-10, 2013.

52. Tutorial: Asset TradeOff Analyst (ATOA), Version 1.0, AgileAssets, Austin, Texas, May 2011.
53. Kim, Seung-Jean and S. Boyd, "Robust Efficient Frontier Analysis with a Separable Uncertainty Model," Information Systems Laboratory, Stanford University, California, November 2007 (sjkim,boyd@stanford.edu).
54. New Jersey DOT, "Road User Cost Manual," 114p, April, 2009.
55. Scheinberg, T and P.C. Anastasopoulos, "Pavement Preservation Programming: A Multi-Year Multiconstraint Optimization Methodology," 89th Annual Meeting of the Transportation Research Boardv 2010.

Part Five

IMPLEMENTATION OF PAVEMENT MANAGEMENT SYSTEMS

26

Steps and Key Components of Implementation

This chapter gives no additional attention to project level pavement management since most agencies have overshadowed project level PMS with the AASHTO MEPDG design method. Thus, discussion of implementation will be limited to network level PMS. In the last several decades, the steps of implementation have changed: by this time, many state DOTs have completed several of the steps once or even twice, some with success and some with less success. The steps of good implementation for most state DOTs, large counties, and cities are as follows:

1. Historically developed an in-house system in the 1980s.
2. Updated or upgraded the original system in the 1990s.
3. Recognized deficiencies and investigated improved systems (some agencies currently start here).
4. Requested detailed information about PMS software from software providers.
5. Requested face-to-face demonstrations of software from select provider(s).
6. Prepared RFP or specifications often including focus on information technology project management. Specifications are tailored to individual agency needs.

7. Considered possible ties of PMS to other systems such as MMS and BMS to work toward a unified AMS. Generally minimizing project level at this point.
8. Looked at data collection and data base needs, emphasizing that a good linear referencing system (LRS) is critical to success.
9. Examined where agency stands with GIS. All pavement features now related to GIS. Most agencies have at least mapping capabilities.
10. Supplemented and individualized procurement documents to specifically cover agency needs.
11. Obtained support and buy-in for change from existing staff before replacing an old existing system because education and training are required.
12. Procured the correct software.
13. Implemented and continued training.
14. Monitored and upgraded over time.
15. Continued training.

26.1   Recognize Need for Change

Every agency has individuals who resist change. The best solution to this problem is to have support from the highest possible administrative level in the agency, which makes gaining support from resistors much easier. Even so, if a legacy PMS is in-place, many older staff members may resist accepting and implementing a new system to replace the old system.

Many agencies now are more successful in implementing new or improved PMS software by enlisting the subdivisions in their agencies responsible for planning, programming, budgeting, and maintaining the highway system. The authors know of at least two major state agencies where the impetus for implementing new PMS came from maintenance or programming rather than from pavement section personnel. Pavement management is now a part of broader asset management issues and must be implemented as such. More details will be discussed in Chapter 29.

26.2   User Interface Design/User Experience Design

While the benefits of PMS have long been known and understood by a select few, implementation has been much slower than anticipated or desired.

PMS developers and researchers often blame this on resistance to change and on the fact that DOT personnel are in a "rut." But software developers have come to better accommodate the user's needs and experience. At least one PMS software provider has accepted this concept and undertaken UI/UX design in its PMS [1]. It is not known who else uses this concept, but it does explain why several state DOTs are progressing rapidly up the PMS maturity ladder to state-of-the-art. User interface (UI Design) and User Experience (UX Design) are known jointly as UI/UX design. It does not involve simply making things look pretty, improving graphic design, or providing warm fuzzies that make the user feel good. It does involve: 1) making things for DOT people, 2) realizing those people aren't you, 3) finding out what those people really need, and 4) building the most useful and useable software for people to meet their needs. UI/UX design focuses on:

1. Iterative development of functionality
2. Rapid and flexible response to change
3. Rapidly delivering value to customers

Maynard [1] quotes Anders Ramsey as follows: "Making great software quickly, it turns out, requires collaborating really really effectively with those pesky non-binary entities called people." Understanding this idea simplifies the development of good PMS software and the implementation of good pavement management. User-centered design involves five layers from the bottom up: strategy, scope, structure, skeleton, and surface as defined in Figure 26.1. Each layer drives what happens on the layers above it. If

What goes into making a software product?

- Surface: The visual layer of the product

- Skeleton: Placement of tabs, buttons, text, tables, etc.

- Structure: The way in which features and functions fit together

- Scope: The features and functions that are included in the product

- Strategy: When the business and its customers want to get out of the product.

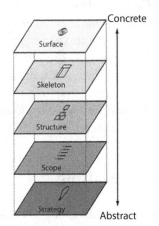

From Jesse James Garrett's *The Elements of Design: User-centered Design for the Web:* http://www.jjg.net/elements/

Figure 26.1 The five "S's" of making good PMS software. After [1]

something is broken on a bottom layer, everything above that layer is broken too. Maynard provides more details and explains difficulties in implementing pavement management over the years.

The authors have long recognized the gains accrued by agencies such as Kansas DOT, North Carolina DOT, and others that have a computer programmer or systems analyst on their PMS staff that understands the eccentricities of pavement management within their agencies. They developed their PMS more actively, innovatively, and beneficially than states lacking such specialists. Clearly a competent computer staff (IT) is essential to the successful implementation of PMS as much as good pavement engineers. Good PMS requires good PM software. UI/UX methods reinforce this effect and recognize that if the software or system is easier to understand, then it will be more successfully implemented by an agency. For example, DOTs without good computer staff might think that the purpose of PMS is mainly to produce pavement condition maps and show relative information that is familiar to PMS users, which is a major fallacy.

26.3   Education/Training

It cannot be emphasized too strongly how important education/training is for implementing good PMS. Education is mandatory to help the agency balance, adjust, and tailor the pavement management process and software to their personnel and its approach to overall asset management. Education can also help to adjust personnel.

Good implementation is almost always tied to the factors discussed in Part Six of this book. A review of the software capabilities presented there will be useful in your agency implementation. Full PMS implementation includes not only the software but also data collection equipment, budgeting, and staffing. Many agencies have reasonable data collection equipment and some have maintained a reasonable pavement management data base for the last several years. Others that need to improve their data collection activities can refer to Part Two for more detail.

26.4   Staffing

Staffing a PMS group is critical. If an agency obtains the fastest software from the best source, they will still need good personnel to master the training provided with the software and to be prepared to implement it effectively. The procurement of software can take time although not nearly as much time as trying to develop your own software in-house. In large organizations, the information technology group may get involved and have approval authority. It is important to provide software documentation to those groups so they will understand and support the PMS group needs.

Because PMS software is complex and costly, it is important to develop an implementation plan that includes investigation of the approval process that exists in the agency. The steps in the process should be done simultaneously and not sequentially if possible. Every effort should be made to get buy-in from as many groups in the agency as possible. There are examples where large DOTs have purchased software, but during the implementation phase, personnel who have not originally favored purchasing the new software become roadblocks to implementation, thus detracting from the full benefits of PMS.

26.5   Agency Input

If the agency has environmental or sustainability concerns or other hard-to-quantify factors, these should be defined as clearly as possible in specific terms so that they can be dealt with by the PMS software. These concerns must be discussed with the provider as the procurement process begins and should be included in a request for proposal. It can be difficult to define such factors in clear terms that can be evaluated in a proposed bid. But as a minimum the agency should be prepared to interface those issues with the software provider when the contract for software installation is in place.

Good software procurement necessitates the agency to provide input to the development of performance models, decision trees for action, contents and display on dashboards for administers, and selection of regular and special reports to summarize the unique information needed in the agency. Most good software providers can customize reports and dashboards but agency personnel need to work with them closely in this implementation process. A mutually-beneficial partnership with the PMS provider must be maintained.

26.6   Training in Software Use

Training is often included with software purchase as well as on-going maintenance of the software. If training is not included, then it needs to be provided for to ensure adequately training personnel during the implementation process. We cannot overemphasize the importance of training in the implementation plan. There will be a need to balance the work-load among people, adjust their training schedules, and tailor the training to agency needs. These demands must continue to be met not just in the initial training with the provider but over the first year or more of the implementation process. The software contract should provide additional and specialized education and/or training as needed. There will also be a need to integrate data needs for the PMS with data to be obtained from the construction process, construction records, maintenance process, and hopefully from a maintenance management system. These interfaces are discussed in Chapters 28 and 29.

27

Role of Construction

The degree of effectiveness of a PMS is directly related to the construction phase, including the provision of vital information to other phases of the PMS. This was emphasized in the 1997 Canadian Guide [2] and reiterated in the updated Guide [3]. More specifically, the construction of roads and pavements has a direct relationship to planning and programming design, on-going maintenance, and periodic monitoring to verify management objectives of expected performance and cost-effectiveness. It is not within the scope of this chapter to describe construction methods, equipment, materials, and environmental considerations; a vast body of knowledge and practice on those topics can be found in various associations, manuals, books, and agency guides and manuals. Rather, this chapter identifies those aspects of construction contributing to good pavement management in terms of:

- Linkages with network level planning and programming
- Linkages with project level design and expected life cycle performance
- Linkages with maintenance and evaluation/monitoring
- Information flows from and to construction, including as-build data and the use of technologies such as GPS coordinates for precise locations

- Role of construction in Public-Private-Partnerships (PPPs) for finance, design, build, operate, and maintain road network or link concessions

27.1 Construction Linked to Planning and Programming

Planning and programming at the network level involves both short- and long-term, management of resources, selection of alternatives including rehabilitation and preventive/preservation maintenance, life cycle economic analysis, and scheduling over the program period. Periodic evaluation/monitoring, as subsequently discussed, is essential feedback to planning and programming, and to design and maintenance, in verifying expected service lines and performance. This enables updating of program schedules and cost estimates.

27.2 Construction Linked to Project Level Design and Expected Life Cycle Performance

Good construction is equally as essential as good design. An inadequate quality, for example, on materials and the construction process itself can invalidate what was expected in life cycle performance. At the design stage, functional aspects and standards are in place, quantities and cost estimates are known, environmental assessments have been carried out, and structural design of the pavement has been carried out. These should all be part of the tender/construction bid package. Then, as-built records on actual materials, quality assurance, layer thicknesses, and variances from the design are important as management links to the design. As well, these are important to any changes or updates that might be needed to the estimated life cycle performance at the design stage.

27.3 Construction Linked with Maintenance and Evaluation

Maintenance and periodic monitoring/evaluation are ongoing activities over the service life of a pavement. The preventive/preservation, corrective and rehabilitative treatments, costs of treatments, timing, and operational aspects are related to the quality of construction and the actual as-built pavement structure. Thus, the availability of construction documentation and data in the PMS database is important to maintenance management. It is also important that monitoring/evaluation plans and operations can have access to this documentation and data.

27.4   Information Flows from and to Construction

Data Base Management, as described in Chapter 12, noted that the requirements for good data have not changed, but the technological, economic, and integrated asset management factors that characterize the present state of data base management have changed substantially. Therefore, construction data acquisition—such as digital video and personal laptops use on site, wireless communication and transmission of data, precise location of construction activities, and recorded data through GPS coordinates—has become a vital part of modern construction management. This means that data transmitted to the PMS data base should be simultaneously accessible to the field and to the office and that it can be cross-referenced to other activities or records.

27.5   Role of Construction in Public-Private Partnerships (PPP's)

Public-Private-Partnerships (PPPs) for roads and other infrastructure have become increasingly popular in various countries around the world. Also known as concessions, they often encompass finance, design, build, and operate for extended periods of up to 30 plus years. Whether a road network or single link is involved, a good asset management system is essential for both the concessionaire and the public agency. This means, among other considerations, that long-term warranty and sustainability requirements must be in place [4] and that good construction and maintenance are essential. As well, construction documentation and linkage to the pavement and/or asset management data base are as important as other linkages discussed in previous sections.

28

Role of Maintenance

The effectiveness of a PMS is directly related to the maintenance phase, the same as construction, including the provision of vital information to other phases of the PMS. Good maintenance management and practices are essential to realizing the expectations from planning and programming, design and construction. In essence, maintenance activities and treatments need to be well timed and executed to ensure that at least minimum acceptable levels of serviceability and safety exist, and/or pavement service life is extended.

This chapter does not describe maintenance methods and equipment, materials and environmental considerations, except for identifying various treatments within maintenance categories. The latter, involving definitions, varies between agencies and countries. For example, common terms are routine maintenance and major maintenance, corrective and preventive maintenance, preservation and rehabilitative maintenance. Often the definition is associated with the agency's budget category. A large body of knowledge and practice regarding these topics is available in agency and association manuals and guides and in the literature. Rather than attempting to cover the various definitions or maintenance practices, this chapter identifies those aspects of maintenance contributing to good pavement management, including:

- Linkages to other phases of pavement management and the associated information flows to and from the maintenance phase
- Pavement preservation as a major part of pavement maintenance and PMS
- Maintenance management systems related to PMS

28.1 Maintenance Linked to Other Phases of Pavement Management

The planning and programming phase of a PMS, a network level activity, includes estimates of maintenance treatments types and amounts, such as km of crack sealing each year and the associated costs over the program period (MMS), with the information transmitted to the PMS data base. Obviously, variations from the original estimates are important, particularly when maintenance treatments and costs become excessive.

Similarly, in the design phase, the life cycle cost estimate of a project needs to include annual maintenance treatments and cost estimates. Again, variables as recorded by the MMS as a section specific activity may well require design updates and the need for earlier than planned rehabilitation. In that aspect, information flows to the PMS data base are particularly important.

The information flows are also important to the construction phase in that earlier or more extensive maintenance treatments than estimated can mean, for example, that there has been a deficiency in the construction quality control or assurance.

28.2 Pavement Preservation in Maintenance

One major change in maintenance has been the rise of the concept of pavement preservation. In one sense, the concept does not change anything that has always been a part of pavement management. On the other hand, it defines maintenance activities in useful ways, usually preventive maintenance required in the early stages of pavement performance life. The National Center for Pavement Preservation (NCPP) was established by Michigan State University and FP^2 Inc. to lead collaborative efforts among government, industry, and academia to advance and improve pavement preservation practices through education, research, and outreach [5].

Section 1507 of US Public Law 112-141 "Moving Ahead for Progress in the 21st Century," Act (MAP-21) defines pavement preservation as follows: "The term 'Pavement Preservation Programs and Activities' means programs and activities employing a network level long-term strategy that enhances pavement performance by using an integrated cost effective set of practices that extend pavement life, improve safety, and meet road user expectations." This definition in its broadest sense could include everything from planning, design, and construction through maintenance and rehabilitation. However, this definition already applies to pavement management. In practice, as defined by David Geiger, [6], pavement preservation consists of three activities: preventive maintenance, minor maintenance (nonstructural), and routine maintenance. This definition has been used worldwide since 2000. It has come to include any treatment applied to the pavement that enhances its performance life without changing its structural capacity. It, therefore, lumps together under the term "preservation" the historically defined concepts of routine maintenance, preventive maintenance, and minor rehabilitation (up to 1½ inch or 40mm thick asphalt concrete overlay).

Pavement preservation based on a survey of state DOTs is defined in [7]. Figure 28.1 from that report shows traditional definitions of preventive maintenance, rehabilitation, and reconstruction, subdividing rehabilitation into minor and major components. In some ways, this figure creates confusion by including minor rehabilitation as part of preservation and subdividing the category "rehabilitation." This is usually clarified by agencies with anything less than 1½ inches of overlay being minor

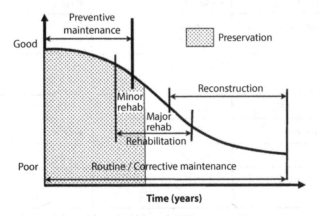

Source: Adapted from Peshkin et al. 2007.

Figure 28.1 Relationship between pavement condition and different categories of pavement treatment. After [7]

rehabilitation (non-structural) and anything over 1½ inches overlay being structural, major rehabilitation. Tables 28.1 through 28.4 define a variety of preservation treatments that are used on hot mix asphalt surfaced and Portland cement concrete surfaced roads. In Tables 28.3 and 28.4 the list is reduced to the most commonly used treatments.

Canada and other countries have also recognized and supported the use of pavement preservation as indicated in [8]. But in general they use the definitions, publications, and ideas presented in [5] and [6].

Table 28.1 Preservation treatments used on high-traffic-volume rural and urban HMA-surfaced roadways. After [7]

Treatment	Treatment Usage	
	Rural (ADT >5,000 vpd)	Urban (ADT >10,000 vpd)
Crack filling	Extensive	Extensive
Crack sealing	Extensive	Extensive
Slurry seal	Limited	Limited
Microsurfacing	Moderate	Moderate
Chip seals	Moderate	Moderate
Ultra-thin bonded wearing course	Moderate	Moderate
Thin HMA overlay	Extensive	Extensive
Cold milling and overlay	Extensive	Extensive
Ultra-thin HMA overlay	Limited	Limited
Hot in-place HMA	Limited	Limited
Cold in-place recycling	Moderate	Moderate
Profile milling	Moderate	Moderate
Ultra-thin white topping	Limited	Limited

Note: Extensive = Use by ≥66% of respondents; Moderate = 33% to 66% usage; Unlimited = <33% usage

Table 28.2 Preservation treatments used on high-traffic-volume rural and urban PCC-surfaced roadways. After [7]

Treatment	Treatment Usage	
	Rural (ADT >5,000 vpd)	**Urban (ADT >10,000 vpd)**
Concrete joint sealing	**Extensive**	**Extensive**
Concrete crack sealing	**Extensive**	**Extensive**
Diamond grinding	**Extensive**	**Extensive**
Diamond grooving	Moderate	**Extensive**
Partial-depth concrete patching	**Extensive**	Moderate
Full-depth concrete patching	**Extensive**	**Extensive**
Dowel bar retrofitting (i.e., load transfer restoration	Moderate	Moderate
Ultra-thin bonded wearing course	Limited	Moderate
Thin HMA overlay	Limited	Moderate

Note: Extensive = Use by ≥66% of respondents; Moderate = 33% to 66% usage; Unlimited = <33% usage

Table 28.3 Preservation treatments commonly used on high-traffic-volume HMA-surfaced roadways. After [7]

Roadway CategoryC	
Rural (ADT > 5,000 vpd)	**Urban (ADT > 10,000 vpd)**
Crack fill	Crack fill
Crack seal	Crack seal
Thin HMA overlay	Cold mill and overlay
Cold mill and overlay	Drainage preservation
Drainage preservation	

Table 28.4 Preservation treatments commonly used on high-traffic-volume PCC-surfaced roadways. After [7]

Roadway Category	
Rural (ADT > 5,000 vpd)	**Urban (ADT > 10,000 vpd)**
Joint seal	Joint seal
Crack seal	Crack seal
Diamond grinding	Diamond grinding
Full-depth patching	Full-depth patching
Partial depth patching	Partial depth patching
Dowel bar retrofitting	Dowel bar retrofitting
	Drainage preservation

28.2.1   The National Center for Pavement Preservation (NCPP)

The National Center for Pavement Preservation (NCPP) [9] was established in 2003 as an independent, quasi-government, non-profit entity devoted to infrastructure preservation. The NCPP is affiliated with Michigan State University (MSU) and sponsored by FP² Inc., which represents the pavement preservation industry.

The NCPP began as a collective vision of nationally recognized practitioners, policymakers, and the beneficiaries of sound pavement and bridge management practices. American Association of State Highway and Transportation Officials (AASHTO), the Federal Highway Administration (FHWA), and FP² Inc. have a common desire to advance the technology and practices of cost effectively preserving pavement and bridge networks in good condition.

The NCPP was established to foster a national advocacy for pavement and bridge preservation at the state and local levels with the focus on national outreach, education, and research in system preservation.

After a decade of service the NCPP has an ambitious program of technology transfer reaching state and local road agencies. NCPP excels in training as its independent position lends non-promotional authority. The key issue for agencies is to have an asset management plan that is successful and provides predictability of good roads. In ten years, NCPP has become a force in pavements. Since most U.S. state DOTs are using PMS, the authors believe NCPP influence would be greater if they defined pavement preservation as part of PMS. In that way the preservation actions used, including cost and treatments, could be recorded into the PMS data base and included in the overall PMS process.

28.3   Maintenance Management Systems Related to PMS

In order to properly program pavement maintenance or preservation, it is important to know the condition of the pavements at that time. A survey of U.S. state DOTs to determine their practices in maintenance condition assessment resulted in a response from 36 of the 50 DOTs [10]. In that context, any agency that has a PMS in place already has a pavement condition assessment program in place.

Results of another questionnaire to state DOTs which captures the practices in maintenance management system (MMS) is presented in [11]. The responding states managed networks ranging in size from 2,500 miles to more than 170,000 miles. Twenty-seven of the responding states said they had an MMS in place. Twenty-three stated that they either had an MMS in place or were planning enhancements of their MMS to include interfaces with other systems such as PMS. Figure 28.2 shows a summary of responses to that survey.

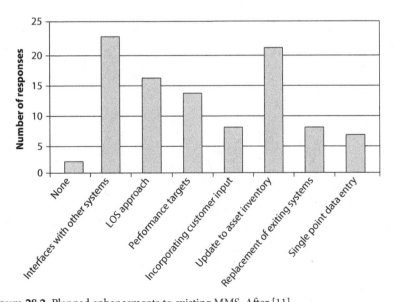

Figure 28.2 Planned enhancements to existing MMS. After [11]

It is beyond the scope of this book to survey software providers to determine which provide both PMS and MMS software. Information on this subject is available in [12]. One provider who supplies both MMS and PMS to state DOTs and other agencies [12,13] shows 14 agencies that use both their PMS and MMS and 15 agencies that use only their MMS. It is obviously more cost effective and easier to interconnect two systems with the same basic system foundation provided by a common vendor.

29

Research Management

Research and/or research and development is directed in general to advancing technology and processes and tracking problems with materials, procedures, environment, economy, and decision making, among many issues. The intent of this chapter is to focus on how research is important and fits into pavement management rather than on specific research needs identified in such endeavors as the Pavement Management Roadmap [14].

Given that the identification of research needs is driven by perceptions, opinion, budgets, and anticipated utilization of results or products, the management levels and functions are both technical and administrative in nature. As well, the returns on investing in research are every bit as complex to determine as are the benefits of pavement management. In any case, the levels of research management are primarily agency specific and a function of size and identified needs.

Pavement research programs should incorporate sustainability through knowledge management, realistic policy objectives, and quantifiable performance indicators [15]. Knowledge generation is considered an asset and should be directly linked to knowledge transfer and the same components as asset management. It has also been suggested that pavement research and the associated management should have key performance indicators with regard to the following [16]:

- Research productivity, impacts, and quality
- Preservation of research infrastructure investment
- Organization, productivity, and efficiency
- Return on research investment
- Provision of education, training, and research
- Sustained partnerships and a clearly defined governance/management structure

29.1   Some Key Elements of Research Management

The following are key elements of research management, not in order of priority:

- Level of funding and resources an agency should commit to research. Guidelines on this are limited, but larger transportation agencies should devote at least 0.5 percent of their capital budget in order for the research function to be meaningful.
- Determining who does the research (e.g., in-house, contracted, and/or some combination) and encouraging innovation. This depends largely on the size; resources, and needs of the agency. Contracted research can draw on special expertise, accelerate projects, and provide objectivity. But in-house research can also facilitate implementation and training. Innovation can be encouraged in various ways including through funded research at universities.
- Encouraging partnerships or alliances between public agencies, universities, industry, and institutes or associations. In general, these can be a win-win for all involved, recognizing that in some cases proprietary technology or information may have to be held secure.
- Disseminating research results through publications, manuals and guides, seminars, courses, and webinars, etc. While all research results should be properly documented, the extent and methods of dissemination are largely functions of the policies and underlying motivations of the agency.

29.2   Issues and Examples

The concepts of using PMS data to help determine areas of need in pavements is valid, but the feedback process to allocate defined research funds is lacking. It is most often used during the direct implementation and upgrade of the PMS because when shortcomings are identified by an individual agency they move to correct them immediately as an integral part of implementation feedback with the software provider or the provider of pavement data collection services. This is an excellent economical and rapid approach to improving pavement management at the individual user level.

Needed improvements may be recognized and made in-house, such as better performance models, better or upgraded data collection methods, or improved decision trees. An agency may request that the software provider upgrade or modify their software to provide new reports, better dashboards, alternative programming methods, etc. They can also identify needed improved interfaces with MMS, BMS, Safety Management, and other systems including shared data, linear referencing systems, and combined data bases. Most software providers update software releases at least twice a year to add updates requested by their User Group. This is often more efficient than an individual agency identifying a problem by preparing a problem statement.

Broader multistate needs in the United States are still processed through AASHTO to NCHRP. Only a small portion of allocated defined research funding finds its way to pavement management topics. It seems, for example, that few pavement management research priorities defined by the FHWA Pavement Management Roadmap have yet been funded.

Most pavement research funds in the last two decades have been directed to two major national efforts: Long Term Pavement Performance (LTPP) and development of the MEPDG. These have been followed by state-by-state calibration activities. The results of the MEPDG funded activities are covered in Part Four herein. The benefits of funds expended on LTPP are yet to be fully realized.

PMS feedback data bases are being widely used to calibrate local national models and concepts such as MEPDG and Superpave. One of the only true ways to show that the concepts and design methods produce correct results is to compare their predictions to observed pavement performance over 10, 15, or 20 years. Validation that MEPDG or Superpave produce correct predictions of performance 20 years hence is still to be determined.

Some funds have been used to study social concepts like sustainable pavements and environmental friendly concepts like warm asphalt pavement. So far, however, no great breakthroughs have occurred in these complex, hard to define areas. Nevertheless, the results of the iterative research process has resulted in many significant improvements in the pavement management process as shown in Part Six, and many millions of dollars are being saved by highway agencies all over the world with these improved pavement management systems.

References for Part Five

1. Maynard, Keith, UI/UX Designer, "Innovating the Interface to the AgileAssets Suite," a presentation at the International User's Conference, 2013.
2. Transportation Association of Canada, "Pavement Design and Management Guide," TAC, Ottawa, Canada, 1997.
3. Transportation Association of Canada, "Pavement Asset Design and Management Guide," TAC, Ottawa, Canada, December 2013.
4. Haas, Ralph, L. Cowe Falls and C. Queiroz, "Long-Term Warranty and Sustainability Requirements for Network Level PPP's," *Proceedings*, 1st Int. Conf. on Public Private Partnerships, Dalian, China, August 2013.
5. National Center for Pavement Preservation (NCPP), "At the Crossroads – Preserving Our Highway Investment," Exhibit 2, Michigan State University, 2007.
6. Federal Highway Administration, "ACTION Pavement Preservation Definitions," Memorandum from David R. Geiger, Reference HIAM-20, Washington, D.C., September 12, 2005.
7. Peshkin, D., K.L. Smith, A. Wolters, J. Krstulovich, "Guidelines for the Preservation of High-Traffic-volume Roadways," SHRP 2 Report S2-R26-RR-2, Transportation Research Board, Washington, D.C., 2011.
8. Ontario Hot Mix Producers Association, "The ABC's of Pavement Preservation," published by the Ontario Hot Mix Producers Association, Ontario, Canada, 2014.
9. Personal communication from NCPP, Larry Galehouse Director, March 2014.
10. Zimmerman, K.A. and M. Stivers, "A Model Guide for Condition Assessment Systems," Final Report, NCHRP Project 20-07, Task 206, October 2007.
11. Applied Pavement Technologies, Inc., "The Use of Highway Maintenance Management Systems in State Highway Agencies, APT, Urbana, Illinois, 2005.
12. Uddin, W, W.R. Hudson, and R. Haas, *Public Infrastructure Asset Management*, Second Edition, McGraw Hill Education, New York, 2013.
13. AgileAssets, Inc., "Maintenance Manager™, Maximize your Budget and ROI with the Maintenance Management System that Ensures that the Right Activity is Performed on the Right Asset at the Right Time," Austin, Texas, 2013.
14. Zimmerman, K.A., L. Pierce and J. Krstulovich, "A 10-year Roadmap for Pavement Management," FHWA Executive Report #HIF-11-014, December 2010.
15. Tighe, S., C. Haas, R. Haas, and G. Kennepohl, "Sustainability of Pavement Research Programs through Knowledge Management, Realistic Policy Objectives and Quantifiable Performance Indicators," *Proceedings*, Transportation Association of Canada Annual Conference, Toronto, September 2008.
16. Haas, Ralph, "Generating and Implementing Forward Looking Innovations in Pavement Research," *Proceedings*, Transp. Association of Canada Annual Conference, Halifax, September 2010.

Part Six

EXAMPLES OF WORKING SYSTEMS

30

Basic Features of Working Systems

Many new working systems have been developed since the 1990s. The details of earlier systems in [1] serve as valuable introductory material for students and new people in the pavement management field. Most current pavement management software, circa 2014, is proprietary and extremely complex. We treat those systems in as much detail as possible herein.

Project level PMS forms a basis for all pavement management, but most major work in pavement management since 2000 is tied to network level PMS. Unfortunately project level PMS has reverted primarily to a design function in the form of the MEPDG [2], which is treated in detail in Part Four. This is unfortunate since design alone is not adequate to produce long-lived pavements because of unforeseen and unpredictable variability in inputs and construction as shown in Figure 30.1. In 1963-65, NCHRP-funded major research to solve similar premature failure problems [3].

Figure 30.1 shows the expected performance for a well-designed heavy-duty pavement using a 20-year design period and the actual performance observed for many heavy-duty pavements in the United States during the 1950s and for many pavements around the world today. Field observations often show early deterioration (represented by the triangles), and the extrapolated dashed line shows that the expected life is often less than "designed." Dealing with this type of problem led to the development of modern PMS.

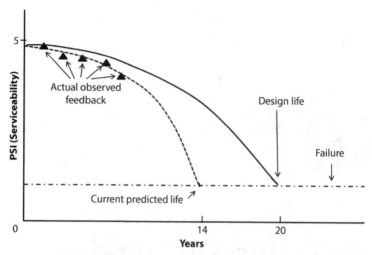

Figure 30.1 The actual observed performance of many heavy duty pavements.

Similar research was carried out in Canada [4–7]. These studies showed that pavement design alone was not adequate to produce required pavement performance. Although good design is certainly the first step in providing desired performance, design also requires accurate predictions of traffic loads and volumes, subgrade strength, base material and surface strength, and as-constructed pavement thickness, as well as surface smoothness. Failure to predict any of these factors accurately can result in early deterioration as shown in Figure 30.1. When the thickness required is greater than actually built or the subgrade is weaker than predicted, then the pavement will also fail early. In the case of the U.S. Interstate Highway System, most early failures were caused by errors in predicting traffic loads accurately or were due to inadequate thickness and materials strength resulting from efforts to get the system constructed as rapidly as possible. Of course, variability can be either a plus or minus, but in real life pavement performance it is usually less than designed because no safety factor is used in pavements such as is used in bridges. However, if the observed variations are "better" than predicted, that will also be taken into account automatically by the PMS to change the dates of future maintenance or overlays. Either way, you win with PMS.

Figure 30.2 illustrates the effect of axle loads on thickness required for a typical design case. In these example conditions, a 25 cm thick surface on weak subgrade should carry 15 million ESALs but a 20 cm thick surface will handle only five million ESALs to failure. The 20 cm surface will likely fail in seven years instead of 20 years if heavier loads exist and thus create

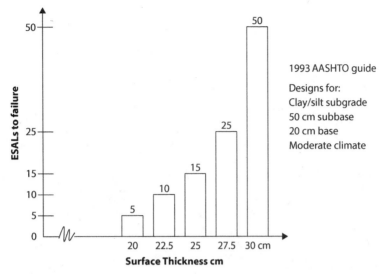

Figure 30.2 Effect of ESALs on Surface Thickness Required.

more ESALs. Therefore, project level pavement management in Chapter 33 is still important.

Despite this obvious reality, pavement management in 2014 has evolved to primarily a network level activity. In response, some current network level PMS working systems have been expanded to include more comprehensive models for determining maintenance and rehabilitation treatments required to account for real variability.

31

Network Level Examples of Pavement Management

Example PMS systems in [1] were the Arizona DOT PMS, circa 1980; the Minnesota PMS, circa 1990; PAVER of the same era; and the MTC regional and municipal system, circa 1986. All of these systems still existed in 2013 but all had changed, becoming more complex, particularly at the state-level where 90% of network level pavement management is done today.

In the 1980s and1990s most PMS was developed in-house or by individual consultants working with a single state such as in Arizona [8–10]. By 1995, there was a significant shift to more uniform commercialized PMS software since the development of software was by then recognized to have complex, unique requirements [6]. "Network Level Examples of PMS" in [1] provide details of the structure of the Minnesota DOT PMS and the Metropolitan Transportation Commission (MTC) PMS as of 1994. Readers should study those details as background since similar details are not available on subsequent systems that are more proprietary. Additional details of the MTC method, now called StreetSaver® are provided in Chapter 33. Due to the proprietary nature of current PMS software developed and used since 2000, in lieu of details not available we will summarize information provided by the predominant providers of PMS software and their users.

Table 31.1 summarizes information concerning the source of PMS in use at that time as reported by the state DOTs to FHWA. The list shows that

Table 31.1 FHWA tally of PMS use in 2002–2003

State	PMS Used	State	PMS Used
Alabama	In house	Montana	
Alaska	Dynatest	Nebraska	In house
Arizona	Stantec	Nevada	In house
Arkansas	Deighton	New Hampshire	Deighton
California	In house	New Jersey	HPMA
Colorado	Deighton	New Mexico	AgileAssets
Connecticut	Unknown	New York	In house
Delaware	AgileAssets	North Carolina	AgileAssets
Dist. of Columbia	Micro-Paver	North Dakota	Deighton
Florida	In house	Ohio	In house
Georgia	In house	Oklahoma	Deighton
Hawaii	In house	Oregon	AgileAssets
Idaho	Unknown	Pennsylvania	In house
Illinois	In house	Puerto Rico	None
Indiana	Deighton	Rhode Island	Deighton
Iowa	Deighton	South Carolina	Stantec
Kansas	KDOT, by URS	South Dakota	Deighton
Kentucky	AgileAssets	Tennessee	Stantec
Louisiana	Deighton	Texas	In house
Maine	Deighton	Utah	Deighton
Maryland	In house	Vermont	Deighton
Massachusetts	Deighton	Virginia	Stantec
Michigan	In house	Washington	In house
Minnesota	Stantec	West Virginia	Deighton
Mississippi	In house	Wisconsin	In house
Missouri	In house	Wyoming	Moving to AgileAssets

18 state DOTS were still using old legacy in-house systems. The three primary suppliers of commercial software at that time were Stantec, Deighton, and AgileAssets.

FHWA has not published or updated such a survey since 2003. Table 31.2 shows a 2011 industry summary of state DOT PMS software use. The number of "in-house" systems had dropped from 18 to 10. At least two other states, Ohio and Mississippi, were in the process of implementing commercial off the shelf (COTS) software as of 2013.

The Tables illustrate the value and growth of the COTS approach and show that Stantec, AgileAssets, and Deighton are the predominant suppliers of PMS software. The Tables also lead one to ask: Which vendor provides the best and most complete software? Since few details of the individual systems are available, it is difficult to answer that question. Stantec had the first state level PMS software in Minnesota [9,11]. Deighton reportedly has the most state level systems in place. But AgileAssets is quickly adding clients, suggesting that they are increasingly highly regarded. Furthermore, AgileAssets provides a complete suite of other systems to couple with PMS, including bridge, maintenance, safety, fleet, facilities, telecom/signals, etc. (Table 31.3).

31.1   Review of COTS PMS Vendors

While it is beyond the scope of this book to evaluate software vendors, two sources have been found who have searched out COTS vendors for a variety of asset types, including pavements [13,14]. We have extracted their summaries for the reader's convenience. The reviews by Brunquet and Mizusawa show that the same three vendors lead the field of state agency PMS in North America: Deighton (www.deighton.com), Stantec (www.stantec.com), and AgileAssets (www.agileassets.com). Uddin has taken information from these websites and has confirmed that these three vendors still dominate the market in 2013 [6].

Brunquet also lists Atlas Exor (www.exorcorp.com), now part of bently .com and EMS-WASP-EMS Solutions (www.ems-solutions.com) as vendors. Of these two agencies, Atlas-EXOR has not penetrated the pavement management market but concentrates instead on broader bridge management and other assets. EMS-WASP-EMS Solutions is primarily an Australian/New Zealand company that has done little work in the pavement area. Thus these two agencies are not considered further in this review.

In 2009, Daisuke Mizusawa completed a report at the University of Delaware with Dr. Sue McNeil for the World Bank, funded by the Japanese

Table 31.2 State PMS industry summary 2011

State	Provider
Alabama Department of Transportation	In-House
Alaska Department of Transportation	Dynatest
Arizona Department of Transportation	Stantec
California Department of Transportation	AgileAssets
Colorado Department of Transportation	Deighton
Connecticut Department of Transportation	Deighton
Delaware Department of Transportation	AgileAssets
Florida Department of Transportation	In-House
Georgia Department of Transportation	In-House
Idaho Transportation Department	AgileAssets
Indiana Department of Transportation	Deighton
Iowa Department of Transportation	Deighton
Kansas Department of Transportation	URS
Kentucky Transportation Cabinet	AgileAssets
Louisiana Department of Transportation and Development	Deighton
Maine Department of Transportation	Deighton
Maryland Department of Transportation	Axiom
Massachusetts Department of Transportation	Deighton
Michigan Department of Transportation	In-House
Ministry of Transportation of Ontario	Stantec
Minnesota Department of Transportation	Stantec
Mississippi Department of Transportation	Deighton
Missouri Department of Transportation	In-House
Montana Department of Transportation	AgileAssets
Nebraska Department of Roads	In-House
Nevada Department of Transportation	In-House
New Hampshire Department of Transportation	Deighton

Table 31.2 State PMS industry summary 2011 (*Continued*)

State	Provider
New Jersey Department of Transportation	Stantec
New Mexico Department of Transportation	AgileAssets
New York State Department of Transportation	Booz-Allen
North Carolina Department of Transportation	AgileAssets
North Dakota Department of Transportation	Deighton
Ohio Department of Transportation	In-House
Oklahoma Department of Transportation	Deighton
Oregon Department of Transportation	AgileAssets
Ottawa, Ontario	Stantec
Rhode Island Department of Transportation	Deighton
South Carolina Department of Transportation	Stantec
South Dakota Department of Transportation	Deighton
Tennessee Department of Transportation	Stantec
Texas Department of Transportation	In-House
Utah Department of Transportation	AgileAssets
Utah Department of Transportation	Deighton
Vermont Department of Transportation	Deighton
Virginia Department of Transportation	AgileAssets
Washington DC District Department of Transportation	MicroPaver
West Virginia Department of Transportation	Deighton
Wisconsin Department of Transportation	In-House
Wyoming Department of Transportation	AgileAssets

Consulting Trust [14]. In addition to AgileAssets, Stantec, and Deighton, they added HDM Global which currently implements HDM-4, an evolving PMS (developed through the World Bank) that includes strong consideration of user costs in its decision process. HDM is treated separately in Chapter 34 herein, but it has not been implemented in any state or provincial DOTs to our knowledge.

Table 31.3 AgileAssets Inc, suite of systems overview. After [12]

Pavement Analyst™ (PMS)	Safety Analyst™
Bridge Analyst™ (BMS)	Sign Manager™
Mobile Apps (Field Data Collection)	Trade-Off Analyst™ (Funding Allocation across assets)
Bridge Inspector™	Signal and ITS Manager™
Mobility Analyst™ (Traffic and Congestion)	Utilities Manager™
Facilities Manager™ (Building Communications, etc.)	System Foundation
Network Manager™	Utilities Analyst™
Fleet & Equipment Manager™	Telecom Manager™
Maintenance Manager™ (MMS)	

We will limit further consideration to the three primary PMS COTS vendors: Deighton, Stantec, and AgileAssets. This should not imply a slight to any other PMS vendors, but these three best illustrate the available PMS software in use as of 2014. It remains possible that new companies will come forward with good PMS software in the future.

31.2   Vendor Background

PMS Limited was begun in 1978 by Drs. Ralph Haas, Matt Karan, and Frank Meyer, who sold to Stanley Engineering in 1981, now known as Stantec. AgileAssets began implementing PMS in 1970 under the name Austin Research Engineers Inc. Deighton began in early 1983 under the leadership of Rick Deighton. After these early starts in PMS, the three firms developed differently. ARE Inc morphed into AgileAssets and now offers pavement management and many other asset management software packages as a part of its asset management suite for states, counties, cities, and private agencies. Deighton started with small implementations in various DOTs in the United States and Canada, generally initiating a simple PMS. They currently report having software used by 17 U.S. DOTs and several Canadian provinces, and they profess implementation in up to 100 other locations worldwide. However, providing PMS support is an ever changing kaleidoscope. Utah, for example, has recently implemented the AgileAssets PMS to supplement their Deighton system because they felt the need for better analysis and the ability to develop better performance models for their pavements. Indiana has adopted AgileAssets Maintenance Management in tandem with its early Deighton PMS. Furthermore, AgileAssets now has PMS in 14 states and provinces and licenses for three more, plus many other locations worldwide (see Table 31.4).

31.3   Guidelines to Available PMS Software

As stated, we do not attempt to evaluate the various PMS software vendors beyond that reported by Brunquet and Mizusawa. We also recognize the risk of perception of "advertising" for the prominent vendors. That is not our intention. We suggest that the reader use the following guidelines to make independent evaluations. The following steps may be useful in such evaluations.

1. Start with the company website and available information but evaluate their claims critically.
2. Search for published technical references from agencies that have implemented the company's software and are actually using it.
3. If this information piques interest, contact at least two or three existing users of the software to discuss their satisfaction with the vendor. Any system that has not been updated in the last three or four years is suspect.
4. The next step is to request that the vendor present a detailed software demonstration at a location appropriate for your needs. Prepare questions in advance and be prepared to penetrate the glossy generalities that vendors often present. You will particularly want to learn information about the details of the software, including the performance models used, how they are obtained, and whether the system uses real optimization as opposed to prioritization or incremental cost benefit (IncBEN) in making its recommendations.
5. You should consider attending the vendor's user group meetings, usually held annually, where you can meet and discuss the software with current users.

Table 31.4 Example: AgileAssets Clients by type of system implemented as of 2013

	System Foundation	MMS	Fleet	PMS	Signs	Signals	BIS	BMS	Safety	TradeOff	Mobility	Net. Mgr	Facilities	Telecom
California DOT – CalTrans	x			x										
Delaware DOT	x			x										
Georgia DOT	x	x		x		x	x	x						
Idaho DOT	x	x	x	x								x		
Indiana DOT	x	x											x	
Kentucky Finance Cabinet	x		x											
Kentucky Transportation Cabinet	x	x	x	x										
Louisiana DOTD & DOA	x	x	L	L										
Montana DOT	x		x	x				L					L	
Montana (State of)	x		x											
New Mexico DOT	x			x										
New York DOT	x	L	L	L			x	L	L	L		L	L	L
North Carolina DOT	x	x		x				x		x				

	x	x	x	L	L	L	L	L	L	L	L	x	L
Ohio DOT	x	x	x									x	
Oklahoma DOT	x	x	x									x	x
Oregon DOT (Inactive)	x			x								x	
Texas CPA (Comptroller)	x		x										
Texas DOT	x	x		x								x	
Utah DOT	x	x	x	x									
Virginia DOT	x			x									
West Virginia Parkway	x	x	x									x	
West Virginia DOT & State	x	x	x			x						x	
Wyoming DOT	x	x	x	x								x	x
Pinellas County, FL - Public Works	x	x		x			x						
Carroll County, FL	x			x									
Newark, DE	x			x									
Montgomery County, DE	x			x									

(Continued)

Table 31.4 Example: AgileAssets Clients by type of system implemented as of 2013. (*Continued*)

	System Foundation	MMS	Fleet	PMS	Signs	Signals	BIS	BMS	Safety	TradeOff	Mobility	Net. Mgr	Facilities	Telecom
Frederic County, DE	x			x										
Wilmington, DE	x			x										
Israel - Maatz	x			x				x	x	x				
Trinidad/Tobago	x	x		x				x						
Quebec	x			x										
Jamaica	x	x												
Dubai	x			x										
Guyana	x	x	x											
Saudi Aramco	x			x										
City of Chicago (Inactive)				x										
# sites	35	17	12	23				5	2	2		1	7	2

6. Do not buy software based on price alone. Some providers offer a skeleton PMS at low cost and then charge extra for adding important subsystems. In general, you get what you pay for in PMS. Proven performance is the key to selecting PMS software.

31.4   Evaluation of Available Information on Leading PMS Providers

31.4.1   Stantec

According to company brochures and their website (www.stantec.com), Stantec was formed in 1950 by Dr. Don Stanley and is a global consulting firm of 11,000 employees with 190 locations in North America plus several international offices which is "one team providing integrated solutions." To broaden their solutions, Stanley Engineering (the forerunner to Stantec) purchased PMS Limited in 1979.

Stantec employs a large suite of engineering, architecture, science, and other technologies with an open enterprise-based architecture and specialized packages for management of individual infrastructure assets including pavements involving, for example, the following functions:

- Data collection
- Condition assessment
- Data processing
- Geographical information systems
- Pavement Asset Performance Modeling and Assessment
- Life-cycle analysis
- Software development and implementation
- Management systems for pavements, bridges, traffic, maintenance, environmental services, and others.

As stated earlier, Stantec implemented modern state network-level PMS software in Minnesota in 1989. Since that time, they have implemented PMS in Arizona, New Jersey, South Carolina, Tennessee, and Virginia. However, Virginia changed in 2010 to AgileAssets. Stantec has also implemented systems in several Canadian provinces, U.S. Federal Land agencies, and municipalities involving over 60,000 kilometers of roads [15].

31.4.2   AgileAssets Inc

AgileAssets is a widely used provider of COTS pavement management software. Starting as Austin Research Engineers in 1970, they morphed into AgileAssets in 1994. In addition to early PMS implementation, they developed the initial concepts of bridge management [16]. As is the case with other vendors, much information can be gained from their website (www.agileassets.com) and from their published literature [17]. Table 31.3 shows the numerous COTS management systems that they provide, including PMS. They list 34 state agencies that currently use their software.

31.4.3   Information from AgileAssets' Clients

As with other PMS providers, AgileAssets provides few details on internal operations of their PMS software beyond information given in Table 31.4. Insight however can be obtained from published literature including [6,18–21] and from users of the system. Idaho Transportation Department employees and AgileAssets staff, for example, describe integration of PMS and MMS [18] as shown in Figure 31.1.

Review of various components of the Idaho PMS software is described by [22–24] and summarized in Figure 31.1 [6] which includes the output from the system to other subsystems, including specifically MMS. Added knowledge of the PMS functions can be seen in Figure 31.2 divided into the following categories:

1. Configuration controls: the way the software uses data contained in the PMS data base.
2. A data base which is the repository of the PMS data collected by various organizational units within the agency.
3. Analysis: the functions used to analyze pavement performance and optimize work programs.
4. Reporting which defines the ways that the system user can present the pavement data and the results of the analyses.

The Idaho PMS uses an integer optimization methodology with multi-constraint and multiyear analysis [24]. More than 13 states in the U.S. are using this integer optimization methodology while other software seems to be using simpler methods such as incremental cost benefit (IncBEN) analysis, general linear optimization, and simple prioritization. Multiyear optimization procedures show strong cost savings [19,20].

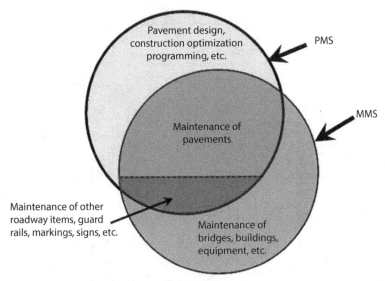

Figure 31.1 Interface of pavement management and maintenance management in Idaho. After [18]

North Carolina DOT has implemented an integrated AgileAssets system comprised of PMS, MMS, BMS, and Asset Trade-Off Analysis (ATOA) [20]. Bhargava presents the framework and applications of the integration method as well as case studies that reveal the positive and stabilizing impact that maintenance has on network condition [19]. In [21] the same team outlines the concept of AgileAssets network safety screening, which is being implemented in the SMS (Safety Management System) for West Virginia DOT. All of these references suggest that AgileAssets provides a sound analytical basis for its PMS software recommended decisions.

31.4.4   Deighton Associates Limited Software, dTIMS-base CT

The best information available from Deighton is provided by their website, [6] and extracted from [25] and two users of their PMS software [26,27]. Deighton has provided pavement management software for over 25 years and considers itself a leader in transportation asset management. They evolved from a small engineering firm providing clients-specific applications for PMS into an international software development organization. According to their literature, they have implemented PMS software in 17 states as shown in Table 35.2. According to Deighton, their clients want to apply the advanced capabilities within dTIM CT, their PMS, to other assets and Deighton reports partnerships with companies involving other

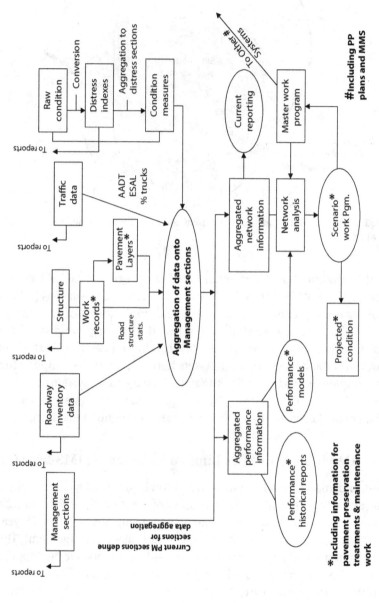

Figure 31.2 PMS system flow framework, as implemented in Idaho. After [22–24]

expertise, although these are not named. They state that their more progressive clients are using dTIM CT to manage not only pavements and bridges but also other data for their networks. These progressive clients are not named. Available literature shows that most of their clients still use Deighton PMS modules only. They report worldwide use of their system by 100 agencies but do not name the agencies.

31.4.5   Information from Deighton Clients

In September 2010, Deighton submitted a report to the Colorado DOT [25] that describes an implementation plan for their system dTIM CT. No subsequent publications have been found which indicate the progress made on that implementation. The report makes it clear that the system was only a PMS which may in the future be expanded, indicating that "as the PMS will form the basis of an Asset Management System (AMS) and is implemented in dTIMS CT already, very little initial work must be done and yearly maintenance will be minor." Thus, while Deighton reports a full-fledged AMS, this statement seems inadequately supported by their own report, and dTIMS CT may not yet be fully developed as an Asset Management System.

Other information on the details and concepts of dTIM CT is provided in [27]. According to that report, the New Zealand Transport Agency (NZTA), formerly Transit New Zealand, with the aid of international consultants has undertaken the implementation of an asset management system in New Zealand. Hatcher points out that the software configuration for dTIM CT "can be undertaken by the owner but Deighton will provide consulting services at an additional cost." In this reference, Hatcher also mistakenly says, "AgileAssets provides hard-coded business logic while dTIM CT can be configured by the owner through consulting services provided by the vendor." AgileAssets has provided PMS software tailored to the client, not hard-coded business logic, for many years. Hatcher also describes a number of missteps and delays that have occurred in New Zealand where a consortium of local agencies and representatives in the transportation agency have tried to do their own implementation. In this sense, he does point out the value of good COTS PMS software properly implemented by people that understand it and its use in lieu of "in-house development and implementation."

In summary, the references by Deighton clients suggest that a critical look at the software is needed before any purchase. Although requested to do so, Deighton did not provide us with further published references from other clients.

31.5   Summary

Clearly many improvements have been made in network level PMS software since 1994. This chapter provides the background information needed for an agency to evaluate and select an appropriate modern COTS PMS.

32

Project Level Examples of PMS Software

As previously mentioned, project level PMS software has fallen out of favor, although it was the original basis and purpose of pavement management. Rather, national research personnel through AASHTO and NCHRP have opted to invest substantial sums in a complex detailed mechanistic-empirical method. This method has had many stops and starts and has been under development from 1995 to 2010 [2].

This minimization of the use of project level PMS in favor of a complex design method is an unusual turn of events given that PMS was developed in the 1970s for the very reason that it was deemed impossible to accurately predict the traffic, climate, materials, and construction variables needed for accurate "design," and furthermore that these variables often changed even from their initial selection. In spite of this history, the intrinsic design orientation of some civil engineers and researchers prevailed with the idea that a more complex and detailed mechanistic design method could be developed and overcome the inherent variability of the problem. During the first two years of research, however, even this group recognized that mechanistic methods alone were impractical and thus they changed the approach to be a mechanistic-empirical design method. The Mechanistic-Empirical Pavement Design Guide (MEPDG) [2], which is treated in Part Four in this supplement, was the result.

Chapter 37 in [1] remains a valuable study tool for pavement engineers and should be used by students and practicing engineers alike. It effectively outlines the life-cycle process required to provide pavements that can properly serve the traveling public for a lifetime. Those desiring to do so can continue to develop project level pavement management by modifying the SAMP software presented in [1] to include any design models they choose, including those from the MEPDG. Although over 350 variables as required in full use of the MEPDG seems unrealistic, the resulting computer time involved may also be prohibitive.

To overcome the lack of use of project level PMS, much of the current network level PMS software is being expanded to produce and use more realistic pavement performance models based on historical data available within the agency's data base. It appears likely that the expanded network approach will continue to be used in the future rather than expanded project level systems.

33

HDM-4 the Upgraded World Bank Model

Currently, Highway Development and Management Tools collectively are referred to as HDM-4, the successor of the World Bank HDM-3. According to Kerali [28], the scope of HDM-4 and its tool set has been broadened beyond traditional project appraisals and is a powerful system for analyzing road management and investment alternatives. New technical relationships have been introduced to model rigid pavement distress, accident costs, traffic congestion, energy consumption, and environmental effects. HDM-4 incorporates three dedicated applications tools for 1) project level analysis, 2) road work programming under constrained budgets, and 3) strategic planning for long term network performance and expenditure needs. Some applications of HDM-4 are presented in Table 33.1, "Change in HDM-4 Management Processes" [29]. While a good deal is said about the improved project level design capabilities in HDM-4, a review of many references including [30–36] shows that only [34] deals specifically with calibrating a pavement deterioration model with HMD-4, in that case rutting. Nearly all of these references deal with the strategic use of HDM-4 and there seems to be little doubt that HDM-4 can be a comprehensive strategic level highway investment analysis tool available to the world.

Table 33.1 Change in HDM-4 management processes. After [29]

Activity	Time Horizon	Responsible Staff	Spatial Coverage	Data Detail	Mode of Computer Operation
Planning	Long term (strategic)	Senior management and policy level	Network-wide	Coarse/ summary	Automatic
Programming	Medium term (tactical)	Middle-level professionals	Network or sub-network		
Preparation	Budget Year	Junior professionals	Scheme level/ sections		
Operations	Immediate/ very short term	Technicians/ sub-professionals	Scheme level/ sub-sections	Fine/ detailed	Interactive

One shortcoming of the HDM-4 is that it typically uses a road matrix obtained from aggregating data about the pavement network from average road data. This differs from typical network level PMS in North American (see Chapter 35 herein) where the precise data for each section in the network is used in the network level or strategic analysis. While a limit to overall precision, the use of matrix level data is understandable in many developing countries. The HDM-4 strategic analysis is useful where good data on the entire network is seldom available. The strategic analysis more completely considers user costs, which are critical in obtaining funding from World Bank and other international lending agencies, than most North American models. Readers desiring a broader strategic study of needs and funding would do well to examine HDM-4 for their use, and a company has been formed to manage and update HDM-4 for all users HDMGlobal (www.hdmglobal.com). It should be noted that HDMGlobal charges a fee for membership, use of the program, and updates obtained there. There is also a suite of applications described in HDM-4 Version 2.05 (http://go.worldbank.org/JGIHXVL460).

Kerali says that "it is essential to note that the accuracy of the predicted pavement performance obtained from HDM-4 depends on the extent of calibration applied to adapt the default HDM-4 models to local conditions." A common theme throughout HDM is that the models used in the analysis require calibration for the country using them. Many of the countries do not have the capability or the data to calibrate those models. As a result, many consultants worldwide are involved in the calibration efforts, as can be seen in the literature. Unfortunately, some apply a one-size-fits-all calibration, or none at all, since there is no way to quantitatively judge results.

33.1   HDM-4 Applications

33.1.1   Functions of HDM-4 within the Management Cycle

In [28], a highway management process can be considered as a cycle of activities that are undertaken within each of the management functions of planning, programming, preparations, and operations. They summarize this approach in Table 33.2. Another reference found on the use of HDM-4 comes from Bangladesh [32] where HDM was used until 1999 after which the HDM-4 became the standard for use by the Roads and Highways Department (RHD), the main user of HDM-4. Professional consultants are not named in the article but financed by the institutional development

Table 33.2 Role of HDM-4 within the management cycle. After [28]

Management Function	Examples of common descriptions	HDM-4 Application
Planning	Strategic analysis system Network planning system Pavement management system	Strategic Analysis
Programming	Program analysis system Pavement management system Budgeting system	Program Analysis
Preparation	Project analysis system Pavement management system Bridge management system Pavement/overlay design system Contract procurement system	Project Analysis
Operations	Project management system Maintenance management system Equipment management system Financial management/accounting system	(Not addressed by HDM-4)

component of DFID have worked since 1994. That article confirms another developing country as a user of HDM-4 and points out that they regularly employ consultants.

Strategy Analysis in HDM-4 deals with entire networks or sub-networks managed by the road organization, for example, the main or trunk road network or a municipal network. Hatcher [28] points out that HDM-4 applies the concept of the road network matrix comprising categories of the road network to be defined according to key attributes that most influence pavement performance and user costs. For example, a road network matrix could be modeled using three traffic categories, two pavement types, and three pavement conditions. This example would apply for one class of roads in the same environmental conditions. Kerali further points out that the quality of the strategic analysis improves as the road matrix becomes more detailed with more categories and sub-categories. Carried to an upper limit, the categories would expand so that individual sections are models. This is, of course, what good PMS software and data does. According to Kerali, strategic analysis may be used to analyze a chosen network as a whole to prepare medium to long range planning estimates of expenditure needs for road development and conservation under different budget scenarios. Estimates are produced of expenditure requirements for medium to long range periods, usually 5 to 40 years [36].

33.1.2 HDM Systems Structure

HDM-4 developers further take the structure of HDM-4 program analysis and strategy analysis to be adequate to handle the first three functions in the management cycle: planning, programming, and preparation. According to them, the application's tool operates on core data objects defined in the following data managers:

1. Roadway Net Manager – defines the road network or sub-network that will be the basis of the analysis.
2. Vehicle Fleet Manager – defines the characteristics of the vehicle fleet that will operate on the road network being analyzed.
3. HDM-4 setup – defines all the default values to be used with the data analysis; a set of default data will be provided with the system but users will have to modify the data to reflect local circumstances.

Figure 33.1 shows a conceptual structure of HDM-4 as defined in [28]. There are four models called technical modules. The first three [Road

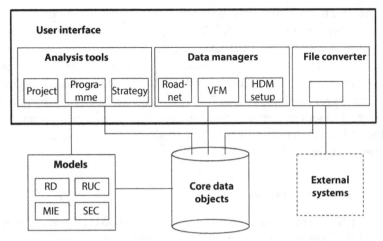

Figure 33.1 Conceptual Structure of HDM-4. After [28]

Deterioration (RD), Maintenance and Improvement Effects (MIE), and Road User Costs (RUC)] are similar in scope to what was originally included in HDM-3. The fourth module, Socio-Economic Costs (SEC), enables the prediction of road accident and environmental impact. According to the authors, these modules are default models provided in HDM-4 and can be substituted by the user or calibrated to local conditions.

The authors make it clear that HDM-4 is full of default models and values and that the user must of necessity calibrate and modify in order to make the program applicable to their situation. History shows that requiring users to calibrate is not simple. Consequently, most agencies in developing countries find it necessary to use a consultant that specializes in HDM-4.

33.1.3   Program Analysis

Program analysis is the middle level of work defined by HDM-4, which is termed Project Selection Level in [1]. The program analysis deals with the prioritization of a defined list of candidate road projects into a one year or multi-year work program under defined budget constraints [1]. This method deals with a list of candidate road projects that are discreet segments within a road network. They say the selection criteria depends on criteria set by the using agencies such as 1) reseal pavement surface at 20% damage, 2) improved thresholds, such as widen roads with volume/capacity greater than 0.8 etc., and 3) development standards such as upgrade gravel roads when average daily traffic exceeds 200 vehicles per day.

It is harder to understand the discussion about the "without project case" versus the "with project case" [28]. The program provides the basis for estimating the economic benefits that would be derived by using each candidate project. The authors suggest doing the program on a multi-year basis but do not define how to accomplish this. They term the second approach as a "multi-year rolling program," which suggests that years beyond the first year are essentially reruns. Such an approach is not a multi-year analysis. Instead, it is a series of single-year analyses that do not consider the effect of work from earlier years.

33.1.4   Project Analysis

According to [28] and others active in developing HDM-4, Project Analysis in HDM-4 is concerned with "evaluation of one or more road projects or investment options." The application analyzes a road link or section with user selected treatments and associated costs and benefits projected annually over the analysis period. Economic indicators are determined for the different investment options. According to the authors, project analysis in HDM-4 may be used to estimate the economic or engineering viability of road investment projects by considering four issues: 1) the structural performance of road pavements, 2) life-cycle predictions of road deterioration, road works affects and costs, 3) road user costs and benefits, and 4) economic comparisons of project alternatives

Pavement management in North America and other areas in the world use basic PMS data for network and project level analysis. This practice indicates that for HDM-4 the agency will use generic data in their strategic and program analysis. Such an approach can create severe problems as was determined in the 1970s when U.S. states were using generic data to report highway conditions to FHWA and the Congress but then used specific measurements in their network analysis. The differences found were so large that they attracted the attention of the U.S. Congress, and FHWA made major changes in the HPMS system [37]. Finally, as stated in [28], "It is important to note that prior to using HDM-4 for the first time in any country, the system should be configured and calibrated for local use."

33.2   Summary

It is concluded in [28] that "the HDM-4 system is seen as the international de facto standard decision support tool for road management." They do not state who "sees" it. Nevertheless, there is no doubt that HDM-4 is better organized and more functional than was HDM-3. It provides a tool for developing countries. In our experience and based on other literature, the most successful uses of HDM-4 are those carried out under the guidance of a well-trained experienced consultant.

34

City and County Pavement Management Systems

A simple Municipal Pavement Design System was described in [1] illustrating such systems. The implementation of similar systems such as StreetSaver® by the Metropolitan Transportation Commission (MTC) which was developed by Dr. Roger Smith, Texas A&M University, Sui Tan MTC engineer, and their team in [6] is summarized herein.

In 1981, at the urging of several San Francisco Bay Area public works directors, MTC conducted a study that identified a shortfall of funding for local road and street maintenance and rehabilitation (M&R) of about $100 million a year for the Bay Area's 17,000 miles of streets and roads. The MTC is the transportation planning, coordinating, and financing agency for the nine-county Bay Area as well as 100 other cities. It functions as both the state-level regional transportation planning agency and the federal-level MPO for 7.3 million residents, 42,600 lane miles of local roads and streets, 23 transit agencies, and a network of seven toll bridges. MTC (http://www.mtc.ca.gov/) is headquartered in Oakland, California, and collaborates with numerous local agency transportation partners in the Bay Area. As the regional MPO, MTC is a steward of federal funds and is accountable for how regional funds, including those for local street and road maintenance, are spent [38].

It was estimated in 1981 that MTC's deferred maintenance needs ranged from $300 million to $500 million. It was determined that the local agencies needed a decision-support system to help manage their road and street M&R needs. The team noted above developed one of the first and best pavement management software programs to assist local agencies with network-level M&R questions. The MTC pavement management program started with six Bay Area community users as a pilot program in 1984. After modifications and adjustments tailored it to the specific needs of cities and counties, the full program got under way in 1986 [39]. The pavement management program and software was later named StreetSaver® and is currently used by all Bay Area cities and over 250 other U.S. agencies, most located in California, Oregon, and Washington. A user group continues to guide improvements and additions to the software and other components.

The success of MTC's PMS program, now the longest-running and best used municipal PMS in the world, is partly, perhaps largely, due to the support MTC provides the users, which includes training, on-call assistance, assistance in addressing budgets, assistance with updating the system, and continuous feedback [39]. The robustness, relative simplicity of the user interface, continuous improvement, and support for the software make it easy for users to apply the software in their pavement management activities. MTC instituted the Pavement Technical Assistance Program (PTAP) in 1998 to help local agencies in the Bay Area implement and maintain their pavement management programs. While implementation was most important early in the process, more recent effort has focused on keeping agency data current and helping agencies prepare reports and presentations that public works personnel can use to demonstrate road and street M&R needs to city councils, county boards, the MPO, and the state. The actual work is provided by consulting firms that have experience in PMS, pavement inspection, and use of the StreetSaver® software. The program is funded with grant funds and supports assistance to Bay Area agencies [39]. It includes a comprehensive data quality management plan [40], and a cadre of consulting firms has arisen that provide pavement management assistance and training to agencies using StreetSaver® outside the Bay Area.

MTC conducted a number of needs studies using the data from StreetSaver® and analysis tools within the software that were presented in the "Bay Area Transportation State of the System 2006" [41]. A local Streets and Roads Working Group was formed of public works personnel in the Bay Area that acts as an advocate for better funding of roads and street M&R needs. They advise MTC on roads and streets funding needs, and with their help and the StreetSaver® pavement management software, MTC was able to develop a regional funding policy for local roads and streets funding that allocates funding not just on need but also on performance. The policy rewards jurisdictions that focus their efforts on pavement preservation rather than on "worst first" funding. Despite the dour economic outlook of recent years, there has been a six-fold increase in that region's investment in local roads and streets over the same period, due in large part to the information generated from analysis using StreetSaver® [38].

34.1   Lisbon, Portugal

In 1999, the city council in Lisbon, Portugal, decided to develop a pavement management system for its road network [42]. The system was the product of cooperative effort among engineers from the University of Coimbra, Portugal; the University of Beira Interior Portugal; and the City Council and Road Conservation Department of Lisbon, Portugal. The basic development took two years and system preliminary implementation began in 2001. It consisted of three basic modules: a road network data base, a quality evaluation tool, and a decision-aid tool which took more than two years to develop. As might be expected, a number of difficulties were encountered in the development. But additional development was planned for the period after 2004. No additional references to the status of the system and the level at which it is used in Lisbon were found for inclusion here.

34.2   City of San Antonio, Texas

In 2009, the City of San Antonio sought pavement management services to include the inspection of approximately 4,000 centerline miles of city streets. Fugro Consultants, Inc. (www.fugro.com) assisted the City with data collection and processing for roadways along with inventory of roadside assets that included sidewalks, curb ramps, drop inlets, signs, manholes, and water valves. Along with the partnership of two subconsultants, ADA compliance inspections were performed on over 50,000 curb ramps on city streets. The pavement condition information was uploaded to the Cartegraph pavement management system (www.cartegraph.com/) and budget analyses for long-term planning were provided to the City. Geodatabases were created for the sidewalk, curb ramp, drop inlet, and sign inventory.

Data collection in February 2010 on this project included 4,800 lane miles and 4,000 centerline miles within the City of San Antonio. This included a single collection pass for every road in the entire network, with arterial and collector roadways being surveyed in both travel directions. High-definition right-of-way imagery, rutting, and roughness data were all collected as were downward pavement images from which surface distress data was extracted. Data was processed to produce Pavement Condition Index (PCI) scores. As an initial part of the processing phase of this project, a sensitivity analysis was performed to determine the impact of specific distresses on the City's condition scores and which distresses most influenced the overall network condition. Based on this evaluation, the remainder of the condition data was processed. Condition scores for the entire network were calculated in the Cartegraph software. This required migration of historic condition data, directly as a condition score, without detailed distresses. Historic scores were used for reference only.

Fugro also collected sign, manhole, sidewalk, ADA ramp, and street width inventory information manually. Fugro provided VisiData (www.visidata.net) web-based software, which allowed the city to review the pavement surface images and right-of-way images from desktop computers. This was done by creating a link from the Cartography software for each segment through a web-browser with the associated VisiData images and summary tables. This allowed City personnel "virtually" to drive any road within the City's network that was part of this survey. The City's shape file was linked to an interactive module within the PMS that facilitates updating and transfer of data between the two. Fugro also trained City personnel on pavement management practices and the technical aspects of operating the PMS software.

This project produced a ten-year network-level budget analysis to project city funding needs to meet their target condition score. Over 30 iterations were required in which funding levels, decision protocols, funding allocation, interest rates, and other variables were adjusted to cover numerous combinations of variables to project required funding to maintain the City roadways. This was a network-level analysis for high-level budget planning. The need for 30 iterations was due to a lack of automated processing available in the Cartegraph software.

Fugro provided a formal report that summarized field activities, network pavement condition, asset inventory and inspection, and budget analyses performed for the City of San Antonio. This final budget analysis was completed in August 2012. City personnel are now operating the Cartegraph PMS and an RFQ was released in October 2013 for the next cycle of pavement condition assessment.

34.3   Metro Nashville PMS Selection Process

In 2008, interviews were conducted with PMS software providers to review and evaluate candidate pavement management systems to develop the Metro Nashville and Davidson County Long Range Strategic Paving Plan [43]. Many candidate systems were rejected early in the process for cost or scope. Systems with a software cost of more than $50,000 were rejected. Other packages were rejected because the vendor required the client to purchase services in order to receive the (often proprietary) software. Still other packages were rejected as insufficient for the size and scope of the Metro road network. Eventually, Cartegraph's PAVEMENTviewPLUS was selected (http://mena.cartegraph.com/solutions/pavementviewplus/) for implementation at Nashville and Davidson County. Items such as ease of use, learning curve, and cost were key factors in the decision making, according to the agency.

PAVEMENTviewPLUS is divided into two modules. The "Segments" module contains inventory data for the network, including current conditions and physical attribute data. The "Segment Analysis Models" module contains the analysis routines and information required to produce a paving plan. Other agencies who have evaluated Cartegraph's PAVEMENTViewPLUS suggest that it is a dressed up; modified version of the MicroPaver protocols [44]. Readers will need to review and judge that for themselves. While inexpensive, PAVEMENTviewPLUS is not well automated.

34.4   Pavement Management in Johannesburg, South Africa

Not all PMS is developed in North America. Much work has been done in South Africa, for example, particularly in the city of Johannesburg which has a road network of some 10,000 kilometers including motorways, major and minor arterials, collectors, formal suburban roads, and roads in informal settlements [45]. The roads are both paved and unpaved. The city implemented its first PMS in 1975 with help from the National Institute for Transport and Road Research (NITRR) Laboratory. Drs. Haas and Hudson both lectured and taught short courses in South Africa at NITRR at about that time, and concepts they presented were incorporated into the pavement management activities in South Africa. A number of upgrades to the city PMS, including a GIS, were made until 2002 when a new system was approved for implementation; the new system introduced the latest PMS technology at that time and concentrated on innovative techniques such as:

- Improving accuracy and completeness of the road map.
- Correcting location of roads in the field which street names were missing.
- Integrating the optimum maintenance and rehabilitation analysis and determination of roads such as motorways and arterials which deteriorate mainly as a result of traffic loading as compared to suburban roads and others which deteriorate mainly as a result of the environment and surface aging.
- Programming for upgrading and maintenance of unpaved roads that can deteriorate overnight as the result of heavy rainfall.

Johannesburg is now a "mega-city" formed in 2001 with an amalgamation of six local councils with the city of Johannesburg. At that same time, Johannesburg Roads Agency was created to manage the city road network under a service agreement with the city of Johannesburg.

The pavement management system:

- Measures the city's road performance by providing the data on key performance indicators.
- Accurately determines the current condition and predicts the future performance of the road network.
- Provides the required information to a city "call center" for Johannesburg—which interacts with the public.
- Determines an annual maintenance and resurfacing program for the street network.
- Determines a required budget to effectively maintain the street and road network.

The pavement management system analyzes paved roads in one of three ways.

1. Using treatments identified from visual distress assessment only, or
2. Using treatments identified from visual distress assessment and verified by an analytical-empirical method, or

3. Using treatments identified from the analytical-empirical method only.

According to the authors, this has resulted in cost saving in the collection of data. Typically, lower volume roads will deteriorate more as a result of surface aging and thus visual distress data only is collected, while on higher traffic roads structural data is also collected for deflection, roughness, and rut depth, since these roads deteriorate as a result of traffic loading combined with aging.

Another feature of the system is the ability to incorporate unsurfaced roads into the PMS. This requires special consideration of pavement condition and scheduling of pavement distress surveys since surface condition of unpaved roads can change overnight due to heavy rainfall and storm water flow. Other city and county agencies that have a combination of paved and unpaved roads would do well to evaluate the work done in Johannesburg for possible information and use in their own agency.

34.5   City of Henderson, Nevada

In 2006, Henderson, Nevada ranked as one of the best 20 places to live in the United States. In 2007, Applied Pavement Technology (APTech) (www.appliedpavement.com/) completed a project for Henderson to develop a road network inventory and extend an existing MicroPaver data base [44]. This effort included PCI inspections of 177 centerline miles of arterial roadways and collector streets as well as 97 parking facilities.

To maintain the transportation infrastructure, reduce the expense of roadway repairs, increase safety for travelers, and enhance the overall quality of life in the City, APTech provided pavement analysis services and pavement management recommendations within budget and a six-month timeframe. Working under the direction of the Henderson Department of Public Works, APTech was responsible for several key activities, including completion of the network definition for all City-maintained arterial streets, collector streets, and parking facilities. Completion of PCI inspections on all the new pavement sections added to the City's network: development of associated MicroPAVER data bases; preparation of geographic information system-based color-coded maps; and on-call technical assistance in other areas related to the pavement management program.

The updated PMS allows the City to efficiently monitor pavement conditions and effectively prioritize pavement maintenance and rehabilitation needs. This completed inventory and condition survey of the city's entire residential street network allows for more fiscally responsible management of the road assets in Henderson.

34.6   GIS Based Pavement Management System—Fountain Hills Arizona

At about this same time, Medina, Flintsch, and Zaniewski developed and implemented a GIS-based PMS in Fountain Hills, Arizona [46]. Research at the University of Arizona produced a case study in which they developed a prototype low-volume roads PMS-based on a GIS platform for the city. For the study, the City Engineer provided the inventory and condition data base already collected and an AutoCAD map of the city. After evaluating several software packages, the team selected the road surface management system (RSMS) package, which had been developed at Arizona State University, as the basic PMS platform. That program was developed to help local Arizona agencies systematically manage low-volume road and street pavements. The researchers evaluated two GIS packages in the study and selected MAPinfo because it was less expensive and easier to learn than other packages. A menu-driven MAPinfo application that runs the RSMS software, imports the pavement maintenance and rehabilitation program, and interactively prepares and displays colored maps with the analysis. The combination of RSMS and MAPinfo significantly reduced the effort required to develop the prototype system, which the city implemented using existing digital data. According to the authors, city engineers were impressed with the prototype system's capabilities. No follow-up published information has been found.

35

Airport Pavement Management

35.1 PAVER and MicroPAVER

In the late 1980s, Dr. Mo Shahin developed a pavement management method with funding through the U.S. Corp of Engineers. His method is called PAVER and has been used for airports, roads, and parking lots [44,47]. Since the method was developed using federal funding, it received a great deal of attention from governmental agencies, such as USDOD [44,48], FAA, and APWA [49,50] for the city and county level. However, no state DOT is known to have adopted the method for use on its highway system.

A simplified version of the method termed, MicroPAVER has been used by cities and in small airport pavement management as shown in the previously cited references. Dr. Shahin reported in 2002:

> MicroPAVER is currently used as the airport pavement management system for airports worldwide from O'Hare International Airport in Chicago to Inchon International Airport in South Korea. Some states use MicroPAVER to manage their general aviation airports including Arizona, California, Colorado, Washington, et al. Furthermore, the U.S. Air Force, U.S. Army, and the U.S. Navy use MicroPAVER to manage their airfield pavements.

Dr. Shahin does not specify which international airports other than O'Hare and Inchon International he refers to, and we could not find references to other such airport use. It is also not clear how many general aviation airports in the states quoted by Dr. Shahin use the method. Whether it is one, a dozen, or all is unclear. There is also no indication whether or not all U.S. Air Force, U.S. Army, and U.S. Navy bases use all or just parts of PAVER such as the PCI; although one reason it is used by the U.S. Army is that it was paid for and developed in the U.S. Corp of Engineers and is mandated.

Nevertheless, it is clear that MicroPAVER is used as an airfield pavement management tool, at least in the United States and possibly in other locations. Two books have been written by Dr. Shahin discussing the method. The first [47] was published in 1994. The book was updated and a second edition was published in 2005, entitled, *Pavement Management for Airports, Roads and Parking Lots* [48]. While the title of the books suggests coverage of a broad range of pavement management activities, it primarily provides background description and a user's manual for PAVER.

PAVER focuses on condition surveys and the development of a pavement condition index (PSI) that ranges from 0 to 100, where 100 is perfect and 0 is extremely bad. The index is widely used even though it is formulated on a non-linear scale, which does not use uniform subcategories for the various categories of good, poor, bad, etc. Significant improvements were reportedly made to MicroPAVER and MicroPAVER in 2002. Version 5.0 was released with several modifications as follows, as quoted by the developers [44].

1. A much improved user interface.
2. Enhanced reporting capabilities.
3. The ability to store additional inventory and condition data.
4. The ability to define new condition indices, which may or may not be based on the pavement condition index (PCI) distresses.
5. The ability to run work plans based on a desired end-condition as opposed to simply a constrained annual budget.
6. The ability to view geographical information system data within MicroPAVER.

35.1.1 Airport Pavement Inventory

MicroPAVER breaks airport pavement inventories into networks, branches, and sections. The network is a group of pavements that are managed together. For example, state aviation agencies manage multiple general aviation airports; consequently, each general aviation airport is defined as a separate network within the state's airport pavement management data base. Commercial and military airports often break airside and landside pavements into separate networks. A branch is an area of pavement that shares a common use. For example, a specific runway may be defined as a branch. A section is defined as a pavement area within the branch that shares similar structural characteristics and loading conditions. A section is also considered to be a management unit, meaning that condition analysis and work planning is performed at the section level then rolled up to the branch and network levels.

According to Shahin, at the request of MicroPAVER users, "user-defined-fields" have been added at the network, branch and section levels to further subdivide their pavement networks. This includes, for example, storing the county location, the latitude and longitude of that airport, and information on funding sources. In general, the 5.0 upgrade purports to simplify data management and data handling for the user.

35.1.2   Airport Pavement Inspection

As a result of the use of PAVER, ASTM developed Standard Practice D5340 "Standard Test Method for Airport Pavement Condition Index Surveys" and D6433 "Standard Test Practice for Roads and Parking Lot Pavement Condition Index Surveys" which are used for performing airside and landside pavement condition inspections. Both standards produce the pavement condition index (PCI) based on visual distress surveys.

35.1.3   Performance Modeling and Condition Analysis

MicroPAVER uses a PCI-based family modeling method. This family modeling procedure is used by grouping the inspection data including pavement age and PCI for similar pavements (similar structure, loading, environmental conditions, etc.) together to generate the deterioration models. Pavement sections are then assigned a family model, but the method does not state how. MicroPAVER is limited to PCI-based family modeling only and does not allow the user to develop other models.

35.1.4   Airport Pavement Work Planning

Work planning in PAVER is categorized as either network or project. Network planning is concerned with "work levels" such as preventive global and major work while project work planning is concerned with "work types" such as overlay thickness, various mix designs, etc. MicroPAVER performs primarily network level planning using the PCI for determining consequences of a fixed known budget.

35.2   USDOT Federal Aviation Administration Support and Use of PMS

In September 2006, the Federal Aviation Administration (FAA) issued an Advisory Circular [51] that outlined pavement management at the project and network levels in details that mirrored almost exactly the details provided by [1]. The Circular outlined the theory behind airport pavement management (APMS), its benefits and components, as well as other background information. The Circular did not give any detail about how to accomplish any of these items, referring rather to the appendices to the advisory circular itself, which can be obtained by the readers if desired. FAA provides a series of such circulars covering skid resistance, nondestructive testing, guidelines and procedures for maintenance of airport pavements, and so forth (www.faa.gov).

The only details about methodology are provided in the Circular, paragraph 4.3, "Sources of PMS Software" [51]. Here they state, "The Micro-PAVER software package may be obtained from an authorized distribution center. There are two such centers, each of which charge an individual fee for distribution and providing updates and corrections." They then refer to the U.S. Corp of Engineers MicroPAVER website (www.usace.army.mil).

Referencing a secondary source, they say, "Other PMS software has been developed and used by consulting engineering firms that provide pavement evaluation and management services. Some firms may offer or sell their software programs for use by an individual or agency." Thus, FAA provides detailed instructions on how to obtain the MicroPAVER software but provides only general information about other possible sources of software. This limits their focus to MicroPAVER.

35.2.1   Detailed Pavement Management Applications

In September 2013, Dr. Mike McNerney, Assistant Manager, Airport Division, AAS-100, Federal Aviation Administration, provided more recent information on airport pavement management [52]. Dr. McNerney has many years of experience as a researcher, private consultant, and now as assistant manager of Airport Engineering in FAA. He was responsible for much of the pavement management developed for Denver and Tampa international airports as well as a number of smaller airports. Based on his experience, he "basically" recommends a three-tiered approach to airport pavement management, as follows:

1. For general aviation airports, with pavement designed for aircraft up to 30,000 pounds (HTTPS:\\FAAPAVEAIR.FAA. GOV\), I recommend the free pavement management system on line using the FAAPAVEAIR software. This is designed for smaller airports, using asphalt concrete pavement, without large loads. Most of these airports are designed for aircraft up to 12,500 pounds.
2. For General Aviation Airports designed for aircraft up to 60,000 pounds with asphalt concrete, I recommend the full MicroPAVER application.
3. For airports with commercial service and loads exceeding 60,000 pounds, I recommend a custom solution that includes distress inspections with additional deflection testing and an engineering approach to pavement management. I feel the MicroPAVER program blindly followed can be misleading or misapplied and decisions made solely on pavement condition index (PCI) can be a poor choice. There are a lot of airports in this category that are using MicroPAVER but should be using a more enhanced pavement management system like the Geospatial Airfield Pavement Evaluation and Management System (GAPEMS) that was used at the Denver International Airport [53,54] and which has saved the airport money by more accurately distributing its maintenance and replacement resources."

Thus Dr. McNerney goes beyond the FAA circular by pointing out that the MicroPAVER is useful for General Aviation Airports but that a broader engineering approach should be used for major air carrier airports such as Denver, Tampa, or Chicago O'Hare.

35.2.2 Implementation of GAPEMS at Denver International Airport

Dr. McNerney is the primary developer of the GAPEMS methodology and applied it at Denver and Tampa international airports before joining FAA. The methodology makes use of GIS and GPS because airports are widely dispersed pavement systems that are not easily defined or located by a linear referencing system (LRS) as are highways. GIS may also be used for highway pavements and city pavement management, but it is especially adapted and needed for airport pavements as shown by McNerney in his significant references [52–54]. Providing more details about the application of GIS to airport pavements is beyond the scope of this book but details may be obtained from these and other references.

35.2.3   Appraisal of other Airport Pavement Management Systems

Michel Gendreau and Patrick Soriano describe the main elements of airport pavement management and review the existing systems in a somewhat-dated 1998 study [55]. Those interested may obtain the article from Elsevier and study it for themselves. Later and more detailed information is provided as follows.

35.2.4   Application of GIS/GPS in Shanghai Airport Pavement Management System

The Shanghai Airport Pavement Management System (SHAPMS) was developed for the Hongqiao and Pudong International Airports in 2005. As reported by Ling and Yuan in 2005 [56], the system was developed to aid airport authorities in determining the most effective application of maintenance and reconstruction work for airport pavements. Since 2005, the SHAPMS has been updated to integrate GIS and GPS technologies to expand the capabilities and utilities of the system [57]. Although the reference does not state specifically, it implies that the details of the PMS used in Shanghai are those provided by Dr. Shahin in PAVER because it references his books and other work [48].

35.3   Arizona Airports Pavement Management System

In 2000, the Arizona Department of Transportation (ADOT) implemented an Airport Pavement Management System (APMS) to monitor the condition of the Arizona Airport Pavement Infrastructure, made up of some 54 airports including Tucson International Airport and Phoenix Sky Harbor International Airport [58]. Prior to the implementation of its APMS, ADOT did not have the objective data needed to determine the validity of pavement funding requests or to prioritize projects when funding levels were insufficient. Furthermore, it did not know whether projects were being requested in a timely manner.

Now every year the ADOT aeronautics group uses the APMS to identify eligible airport pavement maintenance projects that need funding for the upcoming state's five-year airport development program.

In 2010 a major support effort for APMS was undertaken by contractor Applied Pavement Technology (APTech) with assistance from several subcontractors. They completed an update to ADOT's APMS and assessed the pavement condition using the PCI procedure. They termed this methodology as the "industry standard for visually assessing condition" and referred to FAA advisory circulars and ASTM standards.

The ADOT APMS data base contains information on approximately 15.5 million square yards of pavement. The 2010 study showed that the area-weighted PCI value for the entire network was 75, which was a substantial decrease from the area-weighted PCI of 81 in 2006. The APTech study predicted that a steady decline of PCI to a level of 60 in 2018 would occur if no funding were provided for pavement, maintenance, and rehabilitation. The report presents an APP plan showing that approximately $141.5 million would be needed over the next eight years to maintain the system at its current level based on PAVER and enhancements provided by APTech.

35.4   Washington State Airport Pavement Management System

In 2005, Washington State Department of Transportation (WSDOT) developed a statewide airport pavement management program with assistance from Applied Pavement Technology (APTech), Champaign, Illinois. Collaborating with CH2M HILL and CivilTech, APTech updated the WSDOT Aviation, Airport Pavement Management Program (APMP). The principle objective of that program is to assess the relative condition of pavements for selected Washington State airports in the WSASP and Federal Aviation National Plan of Integrated Airport System (NPIAS). The program does not include a new air carrier airport facility such as SeaTac, which has the technical capability to manage its own programs. The program is a tool to identify system needs, programming decisions for federal grants, and other funding mechanisms. It also uses pavement inventories to identify necessary maintenance, repair, rehabilitation, and reconstruction projects. Since that time, APTech has done several additional projects for Washington State including preparation of a pavement management manual [59]. In 2012, WSDOT Aviation undertook yet another contract with APTech assisted by CH2M HILL and CivilTech to complete an additional pavement study [59]. These studies continue to use the PAVER PCI procedure

APTech's work in Washington State, Arizona, and elsewhere clearly shows that it is a leader in airport pavement management in the United States. Additional information about APTech is available from their website (www.appliedpavement.com/).

35.5   Summary

Airport PMS has definitely expanded since 1994. While other PMS systems are available, PAVER and MicroPAVER are more widely used than others. Readers are encouraged to search for additional details of airport pavement management as necessary to fulfill their needs.

References for Part Six

1. Haas, Ralph, W.R. Hudson and J.P. Zaniewski, *Modern Pavement Management*, Krieger Publishing, Florida, 1994.
2. National Cooperative Highway Research Program, *The Mechanistic-Empirical Pavement Design Guide*, NCHRP 1-37A, Washington D.C., September 2011.
3. Hudson, W.R., F.N. Finn, B.F. McCullough, K. Nair, and B.A. Vallerga, "Systems Approach to Pavement Design, Systems Formulation, Performance Definition and Materials Characterization," Final Report, NCHRP Project 1-10, Materials Research and Development Inc., March 1968.
4. Haas, Ralph and W.R. Hudson, *Pavement Management Systems*, McGrawHill, 1978.
5. Wilkins, E.B., "Outline of a Proposed Management System for the CGRA Pavement Design and Evaluation Committee," *Proceedings*, Canadian Good Roads Association, 1968.
6. Uddin, W., W.R. Hudson, and R.C.G. Haas, *Public Infrastructure Asset Management*, McGraw Hill, 2013.
7. Haas, R.C.G., and B.G. Hutchinson, "A Management System for Highway Pavements," *Proceedings*, Australian Road Research Board, 1970.
8. Kulkarni, R.G., K. Golabi, F. Finn, and E. Alviti, "Development of a Network Optimization System," Woodward-Clyde Consultants, San Francisco, 1980.
9. Pavement Management Systems Limited, "Minnesota's Department of Transportation Pavement Management System: System Documentation," report prepared for MnDOT, 1989.
10. Hill, L. and R.C.G. Haas, "Module E: Multi-Year Prioritization," Developed for FHWA Advanced Course on Pavement Management Systems, June 1990.
11. Haas, Ralph, "Minnesota's Pavement Management Systems: Implementation Recommendations," Report prepared for Minnesota DOT, June 6, 1985.
12. AgileAssets Inc. website, www.agileassets.com, 2013.
13. Brunquet, J., "Comparative Assessment of Road Asset Management Systems Software Packages," University of Waterloo, Canada, July 2007.
14. Mizusawa, Daisuke, "Road Management Commercial Off-the-Shelf Systems Catalog," Version 2.0, University of Delaware, February 2009.
15. Haas, R., K. Helali, A. Abdelhalim and A. Ayed, "Performance Measures for Inter-Agency Comparison of Road Networks Preservation," *Proceedings*, Transportation Association of Canada Annual Conference, Fredericton, New Brunswick, October 2012.
16. Hudson, S.W., R.F. Carmichael, L.O. Moser, and W.R. Hudson, "Bridge Management Systems," *NCHRP Program Report 300*, Transportation Research Board, December 1987.
17. AgileAssets web site www.agileassets.com, accessed December 10, 2012.

18. Hudson, S.W., K. Strauss, et al. "Improving PMS by Simultaneous Integration of MMS," Presented at the 8th International Conference on Managing Pavement Assets, Santiago, Chile, November 2011.

19. Bhargava, A., P. Laumet, C. Pilson, and C. Clemmons, "Using an Integrated Asset Management System in North Carolina for Performance Management, Planning, and Decision Making," Paper for presentation at the 91st Annual Meeting of the Transportation Research Board, Washington, DC, January 2011.

20. Bhargava, A., A. Galenko, and T. Scheinberg, "Asset Management Optimization Models: Model Size Reduction in the Context of Pavement Management System," prepared for presentation at the 92nd Annual Meeting of the Transportation Research Board and publication in the Transportation Research Record, Washington, D.C., January 2013.

21. Azam, Md. Shafiul, Uday Manepalli, and Pascal Laumet, "Network Safety Screening in The Context of Agency Specific Screening Criteria" prepared for the 92nd Annual Meeting of the Transportation Research Board, Washington, D.C., January 2013.

22. Perrone, Eric, "Bootcamp Training - Pavement Management," ITD MAPS Project, Version 1.0, conducted by AgileAssets, Inc for Idaho DOT, June 2010.

23. Pilson, Charles, "Bootcamp Training Manual for Maintenance Management," ITD MAPS Project, Version 1.0, conducted by AgileAssets, Inc for Idaho DOT, updated January 2010.

24. Scheinberg, Tonya and P. C. Anastasopoulos, "Pavement Preservation Programming: a Multi-Year Multiconstraint Optimization Methodology," prepared for presentation at the 89th Annual Meeting of the Transportation Research Board, Washington, D.C., January 2010.

25. Deighton Associates Limited, "Asset Management Implementation Framework for Colorado Department of Transportation," Colorado DOT, Denver, Colorado, September 2010.

26. Hatcher, W. and T.F.P. Henning, "Lessons Learnt: New Zealand National Pavement Performance Model Implementation," paper presented at the 7th International Conference on Managing Pavement Assets, 2008.

27. Hatcher, A. W., "Highway Agency Integrated Asset Management Program Decision Support Tools-Current State of Art," Summary Report, draft document, August 2009.

28. Kerali, H.R., R. Robinson, W.D.O. Paterson," Role of the New HDM-4 in Highway Management," 4th International Conference on Managing Pavements, 1998.

29. Odoki, J.B. and H.R. Kerali, "Analytical Framework and Model Structure," Volume 4, The Highway Development and Management Series, International Study of Highway Development and Management (ISOHDM), Paris, World Roads Association (PIARC), 2000.

30. Bennett, C.R. and W.D.O. Paterson, "Guide to Calibration and Adaptation of HDM," Volume 5, The Highway Development and Management (ISOHDM), Paris: World Roads Association (PIARC), 2000.

31. Archondo-Callao, R., "Applying the HDM-4 Model to Strategic Planning of Road Works," TP-20, The World Bank Group, Washington D.C., 2008.

32. Khan, M.U. and J.B. Odoki, "Establishing Optimal Pavement Maintenance Standards using the HDM-4 Model for Bangladesh," Journal of Civil Engineering (IEB), 38 (1), December 2010. http://jce-ieb.org.bd/pdf-down/3801001.pdf

33. Paterson, W.D.O. and T. Scullion, "Infrastructure and Urban Development Department Report INU77," The World Bank, Washington D.C., 1990.

34. Taniguchi, S and T. Yoshida, "Calibrating HDM-4 Rutting Model on National Highways in Japan," The XXIInd PIARC World Road Congress, World Road Association, Durban, South Africa, 2003.

35. Thube, Dattatraya T., "Highway Development and Management Model (HDM-4): Calibration and Adoption for Low-Volume Roads in Local Conditions," International Journal of Pavement Engineering, Vol 14, No 1, January 2013.

36. Zaabar, I. and K. Chatti, "Calibration of HDM-4 Models for Estimating the Effect of Pavement Roughness on Fuel Consumption for U.S. Conditions," Transportation Research Record: Journal of the Transportation Research Board, ISSN: 0361-1981, Volume 255/2010, p 105-116, September 2010.

37. Federal Highway Administration, Overview of HPMS for FHWA Field Offices, Washington, D.C., April 2003.

38. Romell, T. and S. Tan, "Performance-Based Accountability Using a Pavement Management system, 8th International Conference on Managing Pavement Assets, Santiago, Chile, 2011.

39. Metropolitan Transportation Commission, "Pavement Management Program (PMP) Homepage," Oakland, California, http://www.mtc.ca.gov/services/pmp/, MTA, 2012.

40. Smith, R.E., S. Tan, and C. Chang-Albitres, "Distress Data Collection Quality Management in a Regional Agency," Eight International Conference on Managing Pavement Assets, Santiago, Chile, 2011.

41. "Bay Area Transportation: State of the System 2006," published by Metropolitan Transportation Commission and Caltrans District 4, May 2007.

42. Picado-Santos, L.; A. Ferreira; et al, "Pavement Management System for Lisbon," Proceedings of the ICE – Municipal Engineer, Vol. 157, Issue 3, 2004.

43. Metro Nashville Long-Range Paving Plan, Metropolitan Government of Nashville and Davidson County, Nashville Tennessee 2008-2013.

44. Shahin, M.Y., K. A. Keifer, and J.A. Berkhalter, "Airport Pavement Management: Enhancements to MicroPAVER," presented at the 2002 Federal Aviation Administration Airport Technology Transfer Conference, May 2002.

45. P.A. Olivier, et al., "Advancing Pavement Management Techniques in South Africa's Largest City, Johannesburg," 6th International Conference on Managing Pavements, Brisbane, Australia 2004.

46. Medina, Al., G.W. Flintsch, and J.P. Zaniewski, "Geographic Information Systems-Based Pavement Management System: A Case Study," TRB, Vol. 1652,

Seventh Intl. Conference on Low Volume Roads, December 1999, online date: 2007.

47. Shahin, M.Y., *Pavement Management for Airports, Roads and Parking Lots*, Springer publishing; 1st edition, August 31, 1994.

48. Shahin, M.Y., "*Pavement Management for Airports, Roads, and Parking Lots*, with CD-ROM, ISBN 978-0-387-23465-6, Springer eBooks, 2nd ed. 2005.

49. American Public Works Association, "PAVER: Concrete Distress Manual," APWA, Washington, D.C., 2009.

50. American Public Works Association, "PAVER: Asphalt Distress Manual," APWA, Washington, D.C., 2009.

51. Federal Aviation Administration, "Airport Pavement Management Program," Advisory Circular #AC:150/5380-7A, Washington D.C., September 01, 2006.

52. McNerney, M, personal communication, September 20, 2013.

53. McNerney, M.T. and M.E. Kelly, "The use of TABLETPCS and Geospatial Technologies for Pavement Evaluation and Maintenance at Denver International Airport," presented to the 2007 FAA Worldwide Airport Technology Transfer Conference, Atlantic City, New Jersey, 2007.

54. McNerney, M.T., "Airport Infrastructure Management with Geographical Information Systems: State of the Art," Transportation Research Record 1703, TRB, Washington, D.C., online date 2007.

55. Gendreau, M.l and P. Soriano, "Airport Pavement Management Systems: An Appraisal of Existing Methodologies," Elsevier Transportation Research Part A, "Policy and Practice, Vol 32, Issue 3, April 1998.

56. LING Jian Ming, YUAN Jie, XI Shao bo, et al. "On Development of Shanghai Airport Pavement Management System," Journal of Tongji University (Natural Science), China, August, pp. 1041–1046, 2005.

57. Wenlai Chen, Jie Yuan, Meng Li, "Application of GIS//GPS in Shanghai Airport Pavement Management System," 12th International Workshop on Information and Electronic Engineering (IWIEE)," Procedia Engineering 29, 2012.

58. Applied Pavement Technology Inc., "2010 Arizona Airports Pavement Management System Update," prepared for ADOT Multi-modal Planning Division, Aeronautics Group, Phoenix, Arizona, 2010.

59. Applied Pavement Technology Inc., "Washington Airport Pavement Management Manual," Champaign, Illinois, 2008.

Part Seven
LOOKING AHEAD

36

Analyzing Special Problems

Pavement engineers and administrators always need help with a variety of special problems. A properly developed and active pavement management system can be of great assistance in solving such problems. A number of these special problems were outlined in [1]. The pavement management systems available at that time were not able to solve the most difficult problems. Readers should, however, review the material as background since the problems of that time still exist. In particular, there will always be historical problems of some sort, such as energy issues, alternate sources of materials, new types of pavements and material issues, and changing load conditions. The following sections address current pressing problems.

36.1   Calibration of Pavement Design Methods

A well-functioning pavement management system that has been in use for five or more years can provide much of the information needed to calibrate a new design system like the MEPDG [2]. The MEPDG is covered in Part Four herein, and in that discussion we point out that at least 30 states in the United States and some provinces in Canada are undertaking to "calibrate" the MEPDG to their conditions. This is difficult to do since there are over 350 variables that must be considered in the flexible part. As the method is used, its accuracy can be evaluated more completely long-term using a PMS data base. The data will likely not include traffic spectra since few agencies are likely to have a broad traffic spectra data base for years to come. A file of additional detailed data from design and construction variables used in MEPDG analyses can be stored in a trailer file for those few sections. This individual section data can be developed for variables the designer uses in making decisions. A meta-data file should be set up which defines the accuracy with which the designer feels the inputs are determined; measured, estimated, or defaulted [3,4].

36.2   Superpave Evaluation

The SUPERPAVE volumetric asphalt mix design technology was developed as part of the Strategic Highway Research Program in the late 1980s. Many state/provincial and local agencies in North America, and in other countries, have adopted parts of the procedure. Various manuals and software packages are available, such as SW-2 from the Asphalt Institute (info@asphaltinstitute.org).

Work has been done to show how pavement management data can be used to evaluate the benefit of Superpave [5]. But no real evaluation has been done on how well the method actually performs or at least how much better it is than previous methods. As more information on Superpave mixes becomes available, the amount of cracking, distress, roughness, etc. observed and stored in a PMS can be used along with the Superpave design procedures to evaluate the method over the next decade.

36.3   Warm Mix Asphalts

Another relatively new material concept that is being tried as a result of environmental concerns is the concept of mixing and laying asphalt at reduced temperatures, generally termed "warm mix asphalt" [see (warm-mixasphalt.com) and the National Asphalt Paving Association's 2012 Edition of "Warm Mix Asphalt: Best Practices"]. Historically, asphalt concrete pavements have been compacted at temperatures greater than 300° F. Mixing asphalt and aggregate at these high temperatures results in the emission of hydrocarbon vapors, aerosols, and carbon dioxide into the atmosphere. These high temperatures also cause problems for mixes using recycled asphalt pavement (RAP) since the old asphalt in the RAP tends to burn or carbonize as a result of reheating, thereby causing additional vapors. Experiments with lower temperatures in the vicinity of 250°F show that adequate compaction can be obtained if asphalt modifiers such as zeolites, waxes, or emulsions are added. These have come into widespread use only in the past decade. Often in the past, materials have been used for short periods only to display egregious cracking and distortion after three to four years, which vastly shortens pavement life. If the as-constructed material properties such as density of the warm mix are properly stored in the PMS data base, they can be compared to the observed pavement distresses as a normal part of the pavement management process over a five to ten year period to determine the true performance of warm mixes. The Long-Term Pavement Performance Program (LTPP) is planning to add a program of warm mix asphalt experiments in 2015.

36.4   Corridor Analysis

An important aspect of overall road asset management is the distribution of funds along a particular highway corridor among bridges, pavements, and roadside furniture such as signs, guard rails, etc. Pavement budgets and bridge budgets are often handled separately and there is no accepted way to optimize expenditures between bridges and pavements along a particular highway corridor. In some cases bridges may be repaired for safety reasons and pavements are left in bad shape. As shown in [6], with a good PMS combined with a BMS, it is possible to do a joint analysis along any highway corridor. Their example is an interstate highway corridor in North Carolina. As broader asset management is implemented in state DOTs, this problem should be handled on a regular basis as part of AMS. Another study, using an overall asset index as a cross-optimization tool for pavements and bridges, was carried out for the Province of Ontario [7].

36.5   Improved Pavement Performance Models

The proper functioning of a PMS depends on the accuracy of the performance models used to predict pavement service life and ultimate pavement failure. The shape of the performance curve is needed so that analysis can show when to do major maintenance, minor maintenance, or preventive maintenance to extend the pavement life, or to calculate vehicle operating costs. Currently most software uses straight-line or default pavement performance models. The best software, however, carry out performance modeling as an inherent part of their PMS using the historical distress and roughness data on an individual section or on a class of sections to predict future life and failure mode. The discussion in Chapter 15 provides information on performance modeling, as does Part Four. As well, a study in [8,9] demonstrated a method of developing performance models from historical data.

36.6   Geographic Areas of Heavy Damage

The problem of dealing with areas that exhibit extreme damage due to heavy repetitive loads, floods, or hurricanes plagues highway administrators in many parts of the world. At this writing (2014), states with areas defined by the geologic features of an oil reserve, a coal reserve, or a timber reserve are experiencing extremely heavy damage. An excellent example is the Eagle Ford Shale Area in South Texas [10]. Similar problems exist in West Virginia coal mining areas, in states with heavy timber, and in parts of the United States such as North Dakota where oil shale is currently being developed. Since shale deposits exist in other countries and continents, the problem is spread internationally.

It has long been known, for example, that the Eagle Ford Shale area in Texas contained oil and gas reservoirs but it was impractical to extract them because of the density of the shale. New "fracking" extraction methods developed. They require high pressure pumping of large quantities of sand, water, and chemicals into the shale to fracture it and release gas and oil. Each well requires 20 to 40 truckloads (a total of 80 to 100 tons) of water, sand, and chemicals per day. Many of these loads are carried on farm-to-market or secondary roads designed for loads of 50 tons or less. The result has been destruction of these roads in less than a year. The area bounding the shale deposit, in this case made up of more than 20 Texas counties, could be defined in the PMS as its own subsystem. Additional pavement condition data could be obtained in this area and combined with prior pavement condition data to determine the rate of deterioration and assess the damage against the individual companies involved. Although legislation is pending to provide about $900 million tax dollars toward this problem, no one has yet evaluated how this money should be allocated or how much money is really needed. A good PMS could help highway administrators and legislators to resolve the problem both in this situation and elsewhere.

36.7   Analysis of Heavy Load Corridors

For a number of years, the concept of heavy load corridors has been defined in the United States by NAFTA (North American Free Trade Agreement passed in 1994) where heavily loaded trucks from Mexico and Canada are allowed to traverse highway corridors crossing the United States: Mexico to Canada and Canada to Mexico. Interstate Highway I-35 is the NAFTA corridor. It is possible to analyze these corridors with a PMS by taking intensive traffic data and comparing it to the damage being done. To our knowledge, no such analysis has been carried out to date and the adequacy of funds needed for maintenance and rehabilitation are in question as Highway I-35 has deteriorated significantly since NAFTA was passed.

Two similar examples are 1) the "road trains" across Australia where a tractor unit pulls multiple trailers with total weights in the order of 200 tons, and 2) the movement of freight to individual heavily loaded ports, such as Port of Houston in Texas and several ports in New York, which handle ever-increasing weights as international trade increases. These heavy loads, usually transported in containers, almost always move on the nation's highways even when they ultimately end up on railroad trains. Defining and evaluating these "heavy freight corridors" is possible using a good PMS to determine the additional costs required to establish and maintain the corridors in proper operating condition.

36.8   Summary

The foregoing are merely a few examples of special problems facing highway agencies; these problems could be effectively addressed by a good PMS. Readers can undoubtedly think of many similar situations worldwide.

37

Applications of Expert Systems Technology

In 1994, Knowledge-Based Expert Systems (KBES) appeared ready to rapidly advance the field of pavement management decision making. Haas [1] summarizes KBES systems that were being developed for use in pavements and other areas of civil engineering at that time. As of this writing in 2014, none of these systems are in wide use for pavements. This lack of progress illustrates the difficulty of developing a knowledge base required to define and use a KBES. Instead, traditional PMS has been refined and extended as discussed in Part Six herein.

Haas [1] suggested several promising areas for KBES, including problem solving with engineering judgment and tackling problems that are symbolic in nature. These areas and others still offer potential for expert systems technology. A search of recent literature outlines a system based on KBES, but it requires Dynamic Programming to develop a priority list [11]. Smadi points out that KBES applications in pavement management have been specific and have dealt only with diagnostic aspects at the project level. Smadi examines the feasibility of using KBES for network level pavement management and expanding KBES from diagnostic purposes to performance forecasting.

In 2001, Smadi's concepts were used to develop a simple KBES for pavement management in Mozambique, comparable to the World Bank Model HDM-4 [12]. They developed a condition forecasting model based on limited available data and treatment selection models in concert with Mozambique engineers. They observed that their system compared well with HDM-4 results on heavily loaded roads, but it underestimated the needs on lightly trafficked roads and on the socioeconomic impact.

38

New and Emerging Technologies

38.1 Predicted Advances in PMS

Haas [1] outlines seven general categories of potential "future" technological advances. Since that future is now (2014), the following sections explain how accurately, the authors, were predicting such advances.

38.2 Geographic Information Systems (GIS)

In 1994, it was predicted that GIS would be widely used by 2004. This prediction has been well realized as described in Part Six herein. Nearly all PMSs available in 2014 use GIS and GPS (Global Positioning Systems) to provide unique locations/identification information for pavement sections, bridge locations, and other attributes. They can now also be used to locate specific signs, signals, and other roadside details. GIS functions as input to help define linear referencing systems (LRS) needed to ensure that a large highway network data base is properly described over time with the related pavement data. This is critical to good PMS. One of the best known providers of good LRS technology is Esri® [13].

38.3   New Software, Hardware, Data Bases, and Personal Computers

Predictions in this area have been widely realized. Companies developing computer hardware and software operating systems have spent lots of money to upgrade their technology not only for engineering uses but for all kinds of public infrastructure uses. Cisco Systems, PeopleSoft, and other data base software providers have virtually removed all limitations on data base capacity and manipulation.

38.3.1   Computer Hardware

Hardware has advanced to the point that it is no longer a limitation on PMS implementation. Advances in server technology and associated reductions in price have made it practical to operate large pavement management systems on separate servers, sometimes connected to a central mainframe and increasingly making use of "cloud computing." A number of well known computer companies, such as Dell, Microsoft, IBM, and Samsung, offer a suite of products and services. Web-based access to these products and services is readily available and commonly used.

38.3.2   Personal Computers

Computer advances in the past 20 years have gone far beyond what we foresaw in 1994. Not only have personal computers become smaller, lighter, cheaper, and more powerful, but other personal devices such as smart phones, notebooks, and iPads have developed at an astounding rate. Mobile devices are now widely used in pavement management and their use will continue to grow in the foreseeable future. Part Six illustrates how these devices are an integral part of modern pavement management.

38.4   New Measurement Technologies

A lot of new technology has been developed for PMS use since 1994. Many of the changes are documented in Part Two. Such advances will continue, and while not all advances can be covered in this book, a sampling follows.

38.4.1   Integrated Survey Vehicle

Integrated survey vehicles that travel at highway speeds and can collect data on profile, some distresses, and photo or video images are available from several providers in various countries. Examples are given in Part Two. But surface friction and high speed deflection are still measured with separate equipment. Cost, reliability, and accuracy of data are important considerations for pavement managers. High speed deflections are still to be verified in terms of accuracy.

38.4.2   High Speed Structural Evaluation

For the past 20 years, FHWA has funded the development and refinement of a mobile deflection device or rolling wheel deflectometer (RWD). Several organizations in the U.S. and Europe have developed devices over the past several decades that purport to continuously measure pavement deflections. An excellent history of moving pavement deflection testing devices throughout the world can be found in [14]. That report documents the evolution of both mechanical and laser-based systems from the 1950s to the present.

The more modern versions of moving pavement deflection testing devices include:

- Quest Integrated/Dynatest Consulting Rolling Wheel Deflectometer (RWD).
- Swedish National Road Administration and Swedish National Road and Transport Research Institute Road Deflection Tester (RDT).
- Texas Rolling Dynamic Deflectometer (RDD).
- Danish Road Institute and Greenwood Engineering A/S High Speed Deflectograph (HSD).
- Greenwood Engineering A/S Traffic Speed Deflectometer (TSD)
  - Prototype 1 – Danish Road Directorate TSD (originally called High Speed Deflectograph or Danish HSD)
  - Prototype 2 – United Kingdom Highways Agency TSD (UK TSD)
- Applied Research Associates, Inc. (ARA) Rolling Wheel Deflectometer (RWD) (FHWA).

Opinions differ as to whether one of these devices is "better" than all others. One source answers "not yet" [15]. In practice, however, the RWD and HSD seemed most feasible for implementation but only the HSD is commercially produced [16].

38.4.3   Direct Imaging and Analysis Techniques

It was predicted in [1] that direct imaging of many pavement distresses would become commonplace by 2000–2010. It didn't. What happened is that the capability of image capture has advanced (see Part Two), but the capability of completely automating a system for all distresses, texture, and other elements has not been realized.

38.4.4   Automated Testing Procedures

New test equipment capable of measuring material properties and performing construction quality control is improving but much more refinement is needed. This is evident in the application of the MEPDG, which requires dozens of AASHTO tests and specifications listed in the Guide. The Guide also introduces several new tests that are complex and difficult to reproduce. There is not a large financial incentive for equipment manufacturers to develop new and better testing. Such incentives, if available, will likely require government support and research funding [2].

38.4.5   Interface with Other Systems

Implementing and interfacing two, or up to 11, management systems is a new area where great progress has been made and where momentum will continue. The book *Public Infrastructure Asset Management* [17] discusses developments in this regard. S. Hudson [18] shows how state DOTs are interfacing multiple systems such as PMS, maintenance management, and safety management to move toward broader asset management. There is anecdotal information that several states and local agencies are integrating individual systems such as pavements, bridges, safety, fleet, maintenance, etc. This is the beginning of true public infrastructure asset management [17].

38.4.6   Nanotechnology

Nanotechnology development and applications have substantial potential in road and pavement engineering and management in future decades. While the applications to date have been largely in materials like concrete, Nano sensors and carbon nanotubes are examples of such potential [19]. Although large sums of money are invested in the basics of nanotechnology, application to pavements is still limited [20,21].

38.5   Summary

A lot of progress has been made on the use of new technology since 1994 and continuing developments mean that pavement management will not stand still. The next decade, 2015 to 2025, will offer more technological advancements that pavement managers should welcome. However, there is one major road block: procurement procedures and bureaucracy in public agencies make it difficult to buy or even lease new unproven equipment without preparing substantial specifications and acceptance testing. This hindrance needs to be addressed, perhaps through research and development and a willingness to accept the risk involved.

39

Institutional Issues and Barriers Related to Pavement Management Implementation

39.1   Introduction

The issues addressed in [1] are still largely valid in 2014. Haas [1] stressed that acceptance and successful implementation of PMS requires that institutional issues must be understood and addressed. It also stressed that simply relabeling existing practices as pavement management is not sufficient. It identified several important prerequisites to addressing institutional issues, including the need for decision makers to understand the PMS processes, the need for the PMS to be useable and credible, and the need for ongoing support of the PMS from users.

The issues are classified as: 1) related skepticism, 2) various managerial and organizational concerns and realities, 3) legal and regulatory, and 4) interfacing requirements with other divisions or departments in the agency. The extent of these issues varies from agency to agency and are major obstacles to progress. Further discussion of institutional issues, within the context of a modification of the "Pavement Management Roadmap" [22], is subsequently provided.

A major institutional issue also involves the fact that asset management remains a popular topic, but unfortunately it has not been well

Functional area →	Linear Referencing (LR)		Asset Inventory (AI)		Asset Performance (AP)			Planning (PL)		Operations Management (OP)			Resource Management (RM)			Org. Str. (OS)	System		Security	Reporting
Module	Location based tools	LRS Management	Assets / Inventories	Location based data	Asset condition assessment	Construction History	Models	Analysis Optimization	Planning	Projects / Contracts / Repair Orders	Work Orders	Resource Usage and Accomplishment Recording	Labor Mgmt	Equipment Mgmt	Mat'l Mgmt	Organizational structure	System Config	System Admin	Security	Reporting
Maintenance manager		O	S	X				X	X	X	X	X	O	O	O	O	O	O	O	X
Fleet manager									X	X	X	X	O	O	O	O	O	O	O	X
Facilities manager			F	X					X	X	X	X	O	O	O	O	O	O	O	X
Asset manager			Asset													O	O	O	O	X
Network manager	X	O														O	O	O	O	X
Pavement manager		O			X	X	X	X	X		X					O	O	O	O	X
Local Pavements								X	X							O	O	O	O	X
Bridge manager			B	X		X	X									O	O	O	O	X
Safety manager	X					X	X	X	X		X					O	O	O	O	X
Tradeoff analysis								X	X		X					O	O	O	O	X
FDC				X	X	X					X	X								X
PDA					X						X	X								X
System																O	O	O	O	

O = This functional area needs to be covered just once across all modules.

X = This functional area needs to be covered multiple times - once per module.

Figure 39.1 Functional areas of asset management. After [18]

implemented in most agencies. In fact, it has been implemented only modestly, as pointed out in *Public Infrastructure Asset Management* [17]. That reference describes a broader concept of asset management than can currently be implemented in many agencies. As well, Figure 39.1 [18] summarizes functional areas where progress has been made on asset management.

39.2 Summary

The core pavement management community needs to expand its horizons beyond detailed design methodologies and pavement evaluation technologies and provide better tools that provide broader assistance to administrators, financial officers, and long-range planners. Part Six provides guidance for broader solutions through use of modern PMS software.

40

Cost and Benefits of Pavement Management

40.1   General

The proper use of pavement management results in qualitative and quantitative benefits. Most modern PMS optimize solutions for the available budgets. This approach should produce benefits in terms of savings and improved pavement performance. Other benefits include showing management, legislatures, and the road-using public that their money is being spent wisely. A good PMS will also provide the ability to set up and analyze a subset of a network, such as areas where gigantic efforts are going into oil/gas production expansion, namely the Eagle Ford Shale in Texas and the oil sands in Alberta, Canada (see Chapter 36). Another benefit of a sophisticated PMS is the ability to allocate funds among different assets (pavements, bridges, guardrails, etc.) using a trade-off or cross-optimization analysis [7].

A study of PMS Benefit/Costs [23] defined *Ex post facto* and *Ex ante* concepts. The *Ex post facto* analysis concept is shown in Figure 40.1 and is the type of analysis used earlier in Arizona [24].

The *Ex ante* analysis (Figure 40.2) requires the prediction of pavement performance before and after a PMS is implemented. This *Ex ante*

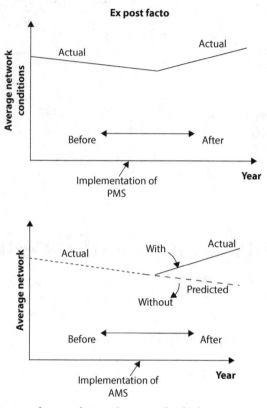

Figure 40.1 Concepts of *ex post facto* evaluation. After [23]

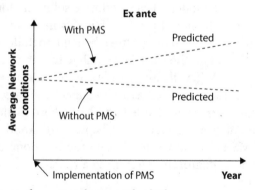

Figure 40.2 Concept of *ex ante* evaluation. After [23]

prediction of pavement performance both with and without pavement management is difficult. Performance indicators to be used for determining benefits are often defined in terms of roughness or serviceability index (PSI) for pavements. In [25], all of these potential factors are tied together

Table 40.1 TAM maturity scale. After [26]

TAM Maturity Level	Description
(1) Initial	No effective support from strategy, processes, or tools. There can be lack of motivation to improve.
(2) Awakening	Recognition of a need, and basic data collection. There is often reliance on heroic effort of individuals.
(3) Structured	Shared understanding, motivation, and coordination. Development of processes and tools.
(4) Proficient	Expectations and accountability drawn from asset management strategy, processes, and tools.
(5) Best Practice	Asset management strategies, processes, and tools are routinely evaluated and improved.

for transportation assets in general. They bring into play the concept of asset management maturation zone or scale, as presented in Table 40.1 [26]. This maturity scale does a good job of evaluating the historical levels of PMS over decades. The ability to analytically assess benefit/cost arrives at Level 4 and Level 5, "Proficient" and "Best Practice" approaches in Table 40.1.

It is now clear after several decades of pavement management usage that the true benefits of PMS increase greatly as an agency system matures from Level 1 to Levels 4 and 5, "Proficient" and "Best Practice."

A standard Benefit/Cost Analysis (BCA) procedure for each of the asset classes under consideration has been put forward in [25]. Example benefit and cost factors are those previously described using the following performance measures:

- Cost measures
  - Data collection costs (equipment, labor, other)
  - Program implementation costs (software, organizational changes, other)
  - System operation costs (additional labor, system/ program maintenance)
- Benefit measures
  - Asset failure and replacement costs (asset value, number of fatalities/injuries, traffic delay hours, labor costs, other costs)

> ○ Time savings on maintenance, administrative tasks, decision-making, and work order placement
> ○ Asset maintenance expenditure savings.

40.2   Quantifiable Benefits

Most of the benefits referred to in Section 40.1 are difficult to quantify. But it has often been reported [1,27] that road authorities can accomplish 5–10% more work with a fixed budget by using a PMS. If the budget is $500 million, for example, at only a 5% level this would result in savings of $25 million. For a cost of operating a PMS roughly at $1–2 million per year, the Benefit/Cost ratio is 25 to 1. At $2 million annual PMS operating cost, the B/C ratio is still 12.5 to 1.

The Texas DOT has reported that tracking performance of all pavement mileage helps know when and how to better expend maintenance funds. They have a $1.4 billion annual maintenance budget. Assuming 40% of that budget, about $600 million, is spent on pavements and that TxDOT feels the use of a management system saves 5% of maintenance funds, an annual saving of $30 million accrues. This represents a B/C ratio of 15–30 to 1, based on the cost of operating a MMS of $1–2 million/year in a large state like Texas. If a DOT wants to better quantify the benefits of PMS, they should also consider other engineering applications of that system [24].

40.3   Benefit/Cost of Developing and Using PMS

In the 1990s, it was difficult to assess financial costs of pavement management software. The costs obviously depend on the size of the agency and highway network being managed and the number of users who have access to the software on a per-user basis. Costs also depend on the customization the agency requires in its contract agreement with the provider. Information based on discussions with various DOTs and PMS software providers suggest that upfront acquisition, installation, and training on software could run from $700,000 to $1.7 million for a DOT with 30,000–50,000 center line miles of highway and 20–50 licensed users. Subsequent annual software maintenance and annual user license fees can run $200,000–$700,000 annually, again depending on size and upgrades.

Similar anecdotal information shows that these agencies spend $300 million to $700 million annually on building and maintaining pavements. The same varied sources show that their pavement budgets can be extended 5-10% further using PMS to provide the correct treatment in the right place at the correct or optimum time. This is in effect a cost savings or benefit of pavement management. Since none of these figures are precise, let's look at a range of possibilities.

Let's consider an example DOT. To be realistically conservative, use an initial or first year cost of $1.0 million to acquire the PMS software and an annual in-house user estimate of $500,000 for staff and operation of the PMS. Carrying the example further, spread the initial cost over three years at an annual cost of ~$330,000. Therefore, average annual PMS costs are $500,000 + $330,000 = $830,000/year. If the annual pavement budget is $500 million, saving of only 1% due to PMS would produce $5.0 million annual saving. Extending this further produces the following:

> 1% savings = $5.0 million – B/C = $5M divided by $0.83 M/yr
>   = 6.0 B/C Ratio
> 2% savings = $10.0 million – B/C = $10M divided by $0.83 M/yr
>   = 12.0 B/C Ratio
> 5% savings = $25.0 million – B/C = $25M divided by $0.83 M/yr
>   = 30.1B/C Ratio

Thus, for even modest expected savings using a good PMS, the benefit/cost ratio ranges from 6 to 30. These calculations do not include the extra benefits obtained in maintenance because of better data furnished to the maintenance section from the PMS. They also do not include benefits to senior administration from up-to-date knowledge of pavement conditions and future needs or the ability to better communicate with legislatures.

40.4   Example Benefits of PMS for Arizona DOT

The Arizona DOT has used PMS since 1980, and in 1998 decided to study its effectiveness [24,28]. Performance indicators in the state (roughness, cracking, etc.) were studied, and a statistical analysis [29] showed roughness (IRI) to be the most significant finding. The analysis found an average initial roughness (1981-1983) to be 68.3 inches per mile (IPM) with a rate of increase of 1.96 IPM per year. The 1993-1995 (post-PMS) period showed average initial roughness of 64 IPM and a rate of increase of only 1.86 IPM per year. Thus on average, pavements were 6.7% smoother after many years of good PMS.

It was also found that pre-PMS pavements on average reached the ADOT maximum tolerable IRI level of 93 IPM at 12.3 years of age. For the after-PMS period (1993-1995), this was 14.9 years. If we compare the extra life benefits to the costs of operating the PMS, we can reasonable estimate a benefit/cost ratios. Overall, the average benefits totaled $423 million and total PMS costs reported by ADOT were approximately $8.3 million in 16 years [24]. This gives an overall benefit/cost ratio higher than 50 to 1.

An improved level of performance also produces savings in user costs. According to World Bank experience [30], user benefits can be four to ten times road expenditures. Even if half the benefits were due to improved materials and construction changes and not PMS implementation, the B/C ratio would still be about 25 to 1.

40.5   Example Benefits of Management Systems for Pinellas County Public Works, Florida

In 2005 the Public Works Department of Pinellas County, Florida, (PCPW) set out to improve benefits in the department by employing best business practices. In October 2006, AgileAssets provided them with a web-based MMS (Maintenance Management System). More than 40 existing computer systems in Pinellas were replaced with three new systems.

In 2011, PCPW reported [31] major cost savings, much greater organizational efficiency, and higher productivity, including the following quantified benefits:

- New systems eliminated the need to acquire two other computer systems budgeted near $500,000.
- The Mowing Department alone saved $1.7 million by a better match between quantity and quality, inventory and methods.
- The labor pool was reduced to 51 employees; there also was a reduction to 70 pieces of equipment.
- Productivity in units per hour increased by 45%.
- In 2004, it had been anticipated that the annual savings produced by the new systems would be $2-3 million, but the documented budget reduction was $6 million.

Other general benefits were:

- Joint participation of senior management, supervisors, and all staff members resulted in a common goal and improved team spirit in the organization.

- Overall, there were improvements in efficiency, decision making, organizational development, accountability, planning, reporting, speed of information gathering and transparency.
- Public Works now accounts for all maintenance work and resources, cost, location, and accomplishment in terms of being fully tracked.

Pinellas County has demonstrated that smaller organizations can also make major savings with appropriate PMS and MMS.

40.6 Summary

There is no denying that the proper use of a good PMS produces substantial benefits. While there is no precise method of defining those benefits, studies show that B/C can range from a lower bound of 10:1 (and much more counting user costs savings) to an upper bound of 25:1 or more, and 100:1 when you add user cost savings, as does the World Bank.

41

Future Direction and Need for Innovation in Pavement Management

Of particular note in [1] was the changing nature or evolution of pavement research, the key issues existing in the 1990s, and an Activity Based or Generic Structure for Pavement Management, which can be directly tied to the asset management framework. Also highlighted are the elements of successful research, the opportunities for innovation, the advancements in pavement management, and finally the future of pavement management in terms of learning from the past. The authors feel that all of Chapter 45 [1] is still relevant today. However, the authors have continued to maintain a strong interest in forward looking opportunities as reflected in [32,33].

41.1 Pavement Management Roadmap

A list of suggestions to the National Pavement Management Conference in Norfolk, Virginia, in 2007 provided the following:

- Technical improvement needs, such as:
  - Longer lasting better quality pavements
  - Seamless interfacing of the strategic, network, and project levels

- o Performance models that separate traffic and environmental effects
- o Making effective use of the Long-Term Pavement Performance (LTPP) data base
- o Establishing data integration protocols
- o Establishing risk exposure procedures in assessing strategy alternatives
- o "Re-integrating" pavement preservation into pavement management
- Economic and life-cycle improvement needs, such as:
  - o Quantifying the benefits of PMS and of component activities like data collection
  - o Very long-term life-cycle analysis protocols
  - o Quantifying the benefits, or extra costs, of varying risk exposure
  - o Incentive programs for improving PMS processes and application in both private sector and Public-Private-Partnership (P3) contracts
- Institutional improvement needs, such as:
  - o Guidelines for knowledge management and succession planning involving people, technology, and information
  - o Overcoming the challenges of institutional inertia (e.g. barriers) to change
  - o Adapting PMS to P3s, particularly in long-term network contracts
  - o Establishing and integrating agency policy objectives with measureable performance indicators and realistic implementation targets.

In 2010, FHWA funded a study of needs for future pavement management innovation, called a Pavement Management Roadmap [22]. FHWA contracted with Applied Pavement Technologies to assemble representatives from several stakeholder groups to participate in regional workshops held respectively in Phoenix, Arizona; Dallas, Texas; and McLean, Virginia. Approximately 30 people participated in each workshop, usually 20 state highway representatives, 2–3 local agency representatives, 2–3 academics, 3–5 from private industry, and 4–5 from FHWA. Hundreds of potential topics were identified and candidate needs were identified in four themes as shown in Tables 41.1 and 41.2. After prioritization in each individual workshop, global prioritization that combined results of the three workshops and a prioritization by individual members were tallied to produce

Table 41.1 Top 10 listing of short-term needs statements by theme [22]

Needs Statement	Description	Funding
Theme 1: Use of Existing Technology and Tools		
Best Practices for Pavement Management	There is a significant need to assemble and prepare a best practices document for the operational and functional aspects of pavement management. This guide will include a broad range of topics, such as benefits and limitations of data collection equipment and procedures, processes for developing and implementing a linear referencing system and addressing data integration issues, guidelines for developing and updating performance modeling, methods for using pavement management to support agency decisions and allocated funds, and methods for communicating pavement management data to stakeholders	$500,000
Development of Pavement Distress Standards	This study will identify distress to be measured, review current state practice, compare state procedures to current AASHTO protocols, identify areas not currently covered by an AASHTO protocol, develop preliminary protocols, conduct webinars or workshops to obtain state buy-in, and finalize the protocol for AASHTO balloting.	$350,000
Development of Improved Methodologies for Evaluating Data Quality	The study's objective is to develop a standard methodology that can be applied to a wide range of pavement condition data to assess quality in terms of accuracy and repeatability. The study results will establish data collection guidelines and evaluate the impact of variability. A product will be the development of guidelines to improve data quality in terms of collection, processing, and reporting.	$350,000

(Continued)

Table 41.1 Top 10 listing of short-term needs statements by theme [22] (*Continued*)

Needs Statement	Description	
Establish and Develop Equipment Calibration Centers and Guidelines	This study will identify potential calibration sites and recommend calibration frequencies and procedures.	$250,000
Theme 2: Institutional and Organizational Issues		
Communicating Pavement Management Information and Benefits	This study will investigate how highway agencies have successfully gained buy-in from decision makers that have led to increased use of pavement management information. The products will include guidelines for making these types of presentations and a collection of effective presentations that can be used as templates.	$250,000
Improving the Skills of Pavement Managers	This initiative will provide guidance to help agencies evaluate the economic/organizational impacts of workforce development. This study will develop training guides, a web clearinghouse for resources, and information on pavement management careers.	$250,000
Theme 3: The Broad Role of Pavement Management		
Development and Use of Effective Performance Measures	Under this study, examples of effective links between strategic and operations performance measures will be conducted, and guidelines on the use of pavement management measures to support strategic initiatives will be developed.	$250,000

Theme 4: New Tools, Methodologies, and Technologies

Development of Automated Condition Data Processing Tools	Improvements to current tools for automating the processing of some measures of pavement evaluation are needed to accelerate the rate at which survey result become available and improve the consistency and reliability of information. Improvements are needed to the processing of surface distress data, GPR, and rutting.	$800,000
Methods to Quantify the Benefits of Pavement Management	This is a synthesis study in which practices in public and private agencies may be explored to determine current practice. The end product is the identification of effective methodologies that can be used to quantify benefits associated with pavement management.	$30,000
Improving Factors Considered in project and Treatment Selection Decisions	The study must provide guidance for addressing agency challenges that influence the use of this information. The product of this research will be a process for evaluating the decision factors used in the pavement management treatment selection process and guidelines for addressing any existing gaps in the criteria.	$250,000

Table 41.2 Top 10 listing of long-term needs statements by theme [22]

Needs Statement	Description	Funding
Theme 1: Use of Existing Technology and Tools		
Methods of Defining and Calculating the Effect of Pavement Preservation Treatments on Pavement Life	This study will quantify the impacts that pavement preservation treatments have on pavement performance, using measured field data from various geographic regions of the country.	$500,000
Investigation into the Risk, Uncertainty, and Variability in Pavement Management Decisions	The objective of this research is to investigate the various forms of variability affecting pavement management recommendations and to develop a process for evaluating its impact and the overall effectiveness of pavement management recommendations. The results are expected to be able to help agencies determine the amount of data needed to provide credible recommendations and to determine what level of risk is considered acceptable, thereby improving levels of accountability and confidence in pavement management.	$350,000
Theme 2: Institutional and Organizational Issues		
Impact of Pavement Management Investment Levels on Benefits	A product of this study is the development of an analysis approach that determines the relationship between funding expenditures, data reliability, and system outputs. Another product will be the development of a methodology for analyzing these relationships.	$350,000

Methods to Promote Pavement Management as a Management Tool	Pavement management's value is not always well understood, especially among executives and elected officials with short-term positions. Public relations is needed to raise the profile of pavement management and communicate its wide-ranging benefits. Research is needed to know how to be most effective with different audiences.	$100,000
Theme 3: The Broad Role of Pavement Management		
Using Pavement Management Data to Support Design Activities	This study will develop a methodology to enhance the sophistication of pavement performance modeling, determine the availability of data fields for both pavement management and pavement design, determine the compatibility of MEPDG and pavement management prediction, enhance DARWin-ME or develop a stand-alone tool, and recommend adjustments to calibrate one or both motels.	$350,000
National Funding Allocations that Account for State Priorities	This study will result in the development of a methodology for comparing pavement performance that accounts for the differences in state priorities and objectives.	$250,000
Theme 4: New Tools, Methodologies, and Technologies		
Performance Models that Consider a Series of Treatments	This study will include a literature search on the pavement performance impacts of a series of treatments; development of a strategy for evaluating treatments in a series, collection of sufficient data from state agencies to develop, analyze, and validate performance curves; and the creation of guidelines on how to develop performance curves for a series of treatments.	$500,000

(Continued)

Table 41.2 Top 10 listing of long-term needs statements by theme [22]

Needs Statement	Description	Funding
Method for Effectively Modeling Structural Condition	This study will quantify the costs/benefits of network-level deflection testing. The researcher will conduct a survey of practice, validate testing with other static devices, determine precision and bias statements, conduct pilot studies, and develop guidelines.	$350,000
Automation of Surface Texture Characteristics	This study will identify surface characteristics that can be identified and quantified using existing high-speed data collection equipment, potential methodologies for quantifying distress, equipment and analysis gaps, and software and equipment modifications.	$500,000
Identifying Strategies for Incorporating Emerging Technologies into the Pavement Management System	The main research objective is to develop a framework for identifying/evaluating the changes that impact pavement management decisions. The framework should be applicable to a wide range of situations and be demonstrated using data provided by state highway agencies. The final product is a set of guidelines for identifying and evaluating factors that influence the recommendations produced by the pavement management system. A clearinghouse for reporting the evaluation of technology may also be a product.	$350,000

the final results. The top 10 short-term and the top 10 long-term needs statements receiving the most support from the 89 participants are listed in Tables 41.1 and 41.2. A required funding level of over $4 million was ascribed to the short-time needs and an additional $3.6 million to the long-term needs. In our opinion, these funding levels are insufficient and, realistically, closer to $30 million is needed. This is the same amount spent to date on the MEPDG, which is a much narrower topic than PMS.

41.2   Consider User Costs and Vehicle Operating Cost in PMS

Nearly all pavement and bridge projects funded by the World Bank consider user costs and recognize that these costs/benefits of good quality assets can be four to ten times larger than the saving in agency costs as discussed in Chapter 40 [34–36].

Such costs must be considered if the true values of asset investments are to be calculated. Yet most North American transportation agencies do not consider the full user benefits of good pavements and bridges, with exceptions like the State of New Jersey. It is important that we learn how to communicate the necessity of adding user costs and vehicle costs to the B/C equation. This is a psychological, administrative, educational, and technical problem that deserves strong funding and consideration.

The reason that this important topic is not listed as high priority in the FHWA Road Map may well be that the group involved was made up primarily of FHWA and State DOT personnel who have not considered the history. This is even more important in 2014 when highway funds are inadequate and there is a great need to raise additional funds. Better understanding of user savings (benefits) could help to make the case for increased funding.

41.3   Needs for Improved Software

Based on the references previously noted, the FHWA Roadmap, and discussions over the past 20 years between the authors and PMS providers, users, and the general public, the following list summarizes the areas of needed innovation with high potential payoff to the pavement and transportation arena.

1. Integrated Asset Management – Perhaps the most obvious need is true integration into asset management at all levels of DOT agencies. This will need to include training and education of agency personnel and administrators and broader global planning and fund allocation beyond pavements.

2. Develop improved software to permit corridor analysis, such as the major interstate highway corridor (IH 35) from the Mexican border to the Canadian border, so that funding needs can be evaluated and properly allocated among pavements, bridges, safety, and roadsides such as signs, guardrails, etc.

3. Benefit/Cost Analysis – Continue to improve software, data collection, and evaluation to provide rigorous benefit/cost analyses and reporting of pavement management results.

4. Develop software to permit multi-year (10–20 years) planning and allocation of pavement resources within the agency assets under multiple constraints.

5. Multi-Constraint – Improve software that permits evaluating and maintaining a desired level of pavement performance (quality) under limited budgets statewide but also while not allowing budgets allocated in any specific district or sub area to fall below required statutory minimums (maintain rural versus urban balance, for example).

6. Interface PMS and MMS – Continue to develop the best possible interface between PMS and MMS to feedback and make full use of allocation of preservation, maintenance, rehabilitation, and reconstruction funding.

7. Build more complete PMS data bases capable of producing or validating improved design and rehabilitation procedures, though not as complex as required for Level 1 of MEPDG which would overpower any practical network level data base in PMS.

8. Broader Level Concerns – Expand educational horizons to high level personnel in DOTs beyond pavements to include planning, long-term predictions, and administrative needs.

9. Compatible Performance Indices – Develop compatible performance indices among bridge management, pavement management, and asset management to allow budget integration. This might involve the use of Utility Theory [37,38].

10. Sustainability – Much is being said about sustainability or sustainable pavements, although it remains unclear to the authors what specifically the term means. This requires a definition of sustainability concepts and integration into the pavement management process. It also requires comparing monetary versus nonmonetary costs and benefits, and a long-term life cycle framework.

11. Green Pavements/Environmental Concerns –Performance indicators for environmental benefits will be needed to summarize any benefits to be gained. But they must also be compared to potential opportunity costs or loss in performance.

12. Risk – The evaluation of risk exposure in pavement construction and maintenance remains a difficult task. Some methods attempt to quantify risk in terms of reliability, such as having a pavement that will perform as expected for 50% reliability, 95% reliability, etc. These remain to be validated with long-term pavement performance observations.

41.4   Forward Looking Opportunities

The FHWA initiative on the "Pavement Management Roadmap" [22] identifies opportunities to advance pavement management. This initiative is supplemented at the international level by the author's perspective and background [39]. Table 41.3 lists three categories of opportunities, along with example issues/challenges and prospects for major advances:

(A) Pavement Data (Needs and Cost-Effectiveness; Collection Technologies; Quality Assurance; Storage and Integration)

(B) Pavement Management (Structural Design and LCAA; Performance Modeling; Treatment Selection; Quantifying Benefits; Decision Support)

(C) Institutional Improvements (Organizational Structure; Location of PMS and AMS; Technology; Skills; Public-Private-Partnerships)

41.5   Motivating Factors and Roadblocks in Advancing Pavement Management

All aspects of transportation including pavement management must look forward in order to advance. When promoting growth, it is beneficial to

Table 41.3 Forward looking opportunities for advances in pavement engineering and management. After [39]

Forward Looking Opportunity Areas	Example Issues/Challenges	Prospects for Major Advances (Short Term 1-5 Yrs; Med. Term 6–10 Yrs; Long Term 10 Yrs. Plus)
A. Pavement Data 1. Needs and Cost-Effectiveness (comprehensive protocols/guidelines for types of data required, frequency of collection, level (strategic, network or project), use (MEPDG, overall asset management, etc.)	• Responding to advancements in technology • Consistency over time • Amount (s) and types required for different uses • Value of and compatibility with historical data • Role of standardization and comparison across agencies • In-house or outsourcing data collection? • Commitment to and amount of resources required • Coordination with other data collection (traffic, etc.) • Optimization of data collection	• Needs will remain short to long term • More comprehensive guidelines likely in short term • Advances will be constrained by cost concerns short to long term • Widespread standardization not likely in short to medium term • B/C analysis for pavement data will become more established
2. Collection Technologies (precision required, automation of condition measurement, sensors, image quality, speed,	• Implementation of high speed deflection (eg. Rolling Wheel Deflectometer) • Evaluation of new/improved equipment • Equipment costs, reliability and service life	• Cost may delay under spread use of high speed deflection in short to medium term

referencing, integrated collection capabilities, equipment reliability and robustness)	• Degree of integration capabilities required in a vehicle vs. use/optimization of data collection • Agency procurement of equipment vs. contract/outsourcing	• Competition will continue to drive advances short to long term • Image quality will continue to improve (eg. 3D photogrammetry) short to long term
3. Quality Assurance (validity, consistency, accuracy, completeness, management of data quality, audits, effect of collection method, automation of quality checks, QC and QA plans	• Level of accuracy needed for various data elements? • Volume of low-quality data vs less but higher quality data? • Technical expertise required to develop QC/QA plans • Impact of staff turnover (vendors and clients)? • Impact of data quality on engineering and management decisions?	• QA procedures in LTPP can be used advantageously for short to long term • Prospects for impact of quality level of data on pavement design, maintenance, preservation, etc. likely to be better established in medium term
4. Storage and Integration (storage needs and capacity, methods, file sharing, security, access, updates, queries/retrieval, tying "silos" together, integration mechanisms (reference location, asset value, risk exposure, etc.), cost)	• Limits to storage capacity, offsite backup, purging old and/or redundant data • Reluctance to share information/preservation of "silos"? • Distribution, format and level of reports • Sufficiency of available technology and agency resources to meet storage and integration demands	• Benefits from well designed and managed storage systems and integration platform can be substantial in the short term • A constraint on storage capacity; however, is already a short term problem

(Continued)

Table 41.3 Forward looking opportunities for advances in pavement engineering and management. After [39] (*Continued*)

Forward Looking Opportunity Areas	Example Issues/Challenges	Prospects for Major Advances (Short Term 1-5 Yrs; Med. Term 6–10 Yrs; Long Term 10 Yrs. Plus)
B. Pavement Management		
1. Structural Design and LCCA (input variables, type of facility, design method, component models, design options, LCCA method, constructability and maintainability)	• Probabilistic vs deterministic approach • Adoption of the MEPDG method, or …..? • Defining and incorporating sustainability and "green" aspects • In-house vs outsourced design (within a P3) • Communication of finalized design to other areas (construction, maintenance, etc.)	• More probalistic short term • Extensive calibration on MEPDG short term • Comprehensive attention to sustainability and "green" roads short to medium term • More P3's short to long term
2. Performance Modeling (Direct part of evaluating design options, models), predictions (IRI vs Age, and/or…..), reliability, periodic updating, accuracy, etc.)	• Probalistic vs deterministic basis • Impact of new materials on predictions (warm mixes, etc.) • Groups/families vs individual sections • Calibrating MEPDG performance models • Use of performance models in predicting remaining service life (RSL) – functional and structural	• Continual move to probabilistic short to long term • Continued work on improved accuracy of models, and in MEPDG calibration, short to long term • Advances in RSL protection capability short to medium term

3. Treatment Selection (Fundamental component of a PMS, selection process for network and project, interface with other project elements, sensitivity to timing, safety, constructability and future rehabilitation)	• Flexibility in selection *vs* change upon implementation • Clarifying preservation *vs* preventative *vs* rehabilitative *vs* maintenance treatments • Estimating treatment costs in rapidly changing prices • Types and extent of information needed for selection	• Better, more comprehensive models/processes likely over long term • Integration of preservation and preventive treatments into PMS likely in short term • More emphasis on long term impacts of treatments likely over short to long term
4. Quantifying Benefits (Cost side of people, equipment, data collection, etc. represents the investment; impacts on decisions)	• What benefits and how to quantify? • Demonstrating changes in network condition vis a vis cost side • How to communicate benefits and to whom?	• Increasing demand likely from stakeholders to quantify benefits, short to long term • Improved communication tools also likely, short to long term
5. Decision Support (information needed at all levels for policy, strategic network and project level decisions; optimization approach; feedback)	• Incorporating risk exposure into the decision process • Balancing practicalities with recommendations • Incorporating user costs and benefits? • Transparency of the decision process	• Increasing use of risk analyses in decisions, short to long term • Increased requirement from senior levels to demonstrate value of PMS in decision support, short term

(Continued)

Table 41.3 Forward looking opportunities for advances in pavement engineering and management. After [39] (*Continued*)

Forward Looking Opportunity Areas	Example Issues/Challenges	Prospects for Major Advances (Short Term 1-5 Yrs; Med. Term 6–10 Yrs; Long Term 10 Yrs. Plus)
C. Institutional Improvements 1. Organizational Structure (Centralized *vs* regional decisions; simple (small) *vs* comprehensive (large); use of performance indicators)	• Impact of funding (amount, sources) on organizational structure • Capability of adapting to change (downsizing, asset management, retirements, information, politics, technology, etc.) • Ability to compete for pavement dollars	• Many types of changes will occur, even in short term, and adaption will be crucial to survival of pavement management • Continued erosion of institutional knowledge likely in short to medium term
2. Loction of PMS and AMS (Distinct or combined offices for pavement management and asset management; communication channels	• PMS as a "silo" or component subsystem of AMS? • Rationale for PMS in traditional location (materials, planning, maintenance, etc.) • Pavement preservation in the PMS, or separate budget?	• Smooth interpretation of PMS, BMS, etc. into AMS likely to be a struggle over short to medium term • Risk of losing distinct benefit and features of PMS, short term
3. Technology (State-of-the-art technologies today, and periodically upgraded, for data acquisition and processing, sensors, maintenance, etc.	• Developing and maintaining in-house expertise on a fast moving world of technology • Assessing capabilities and limitations of new technologies • Investment in new or improved technologies	• Effective acquisition, understanding and use of new/improved technologies will continue as a long term need

4. Skills (Experience, teaching and training base; periodic upgrades; technical plus administrative and other skills)	• Determining what skills the "leaders of tomorrow" will need (see TAC Briefing Note of Nov., 2009, Ref. 8) • In-house skills/knowledge requirements vis a vis outsourced/purchased skills • Losses through retirements and resignations	• Maintaining the continuing skill sets requirements for effective PMS and AMS will continue at a long term priority need
5. Public–Private-Partnerships (Use ranges from maintenance out-sourcing to finance, design, build and operate)	• Achieving a true partnership with measurable key Performance Indicators on warranties, source delivery, allocation of risk, etc. • Achieving positive benefits for all stakeholders	• Adoption of P3"s in PMS and AMS will be a growing trend over the long term (See Ref. 6)

identify motivating factors as well as roadblocks that must be overcome. The former includes both institutional and individual aspects among the following factors:

- Clear recognition of the challenges remaining to advance PMS.
- Ability to communicate benefits of PMS today as well as new benefits from advances.
- Desire to continually improve practice and add to the state-of-knowledge.
- Willingness to accept risk associated with research, development, and implementation.
- Desire to improve technology, such as performance modelling for maintenance interventions, preservation treatments and rehabilitation, and development of realistic performance indicators for both engineering practice and for stakeholder understanding.
- Recognition of the substantive benefits in moving PMS forward faster through college level intensive training courses.

Roadblocks to advancing pavement management also exist. While not insurmountable, they must be addressed and include the following:

- Institutional inertia in terms of being comfortable with business as usual, plus an aversion to risk and a short term outlook.
- Lack of willingness to commit the necessary resources to research and development. R&D always seems to be vulnerable when economic downturns occur.
- Lack of knowledge of what exists in the literature, either due to turnover of staff and the result of being new to the field, or to an attitude of not willing to study. This is unfortunately a common pervasive roadblock today and presents a substantial drag on advancing PMS and certainly on existing best practice.

42

Developments in Asset Management

Asset management systems have evolved since 1994, and many organizations around the world use AMS for various components of their infrastructure and/or use the asset management system as an umbrella for all components. *Public Infrastructure Asset Management* [17] captures this overall concept and the state-of-practice. As well, a comprehensive package on Transportation Asset Management was prepared for the Alaska DOT & PF in 2010 [39].

A lot of energy has been spent on the concept of asset management (AMS) in the last 25 years. It is generally defined as a top-down process for coordinating all activities related to providing and maintaining the assets of a transportation agency. The earliest organized efforts were taken by FHWA in the 1980s when the U.S. Federal Highway budget included a requirement that all state DOTs implement up to seven management systems, including pavements, bridges, safety, maintenance, and congestion. Within a year, it became evident that this was not a practical requirement and it was rescinded.

The Transportation Research Board began discussion of AMS in 1997 and a formal committee formed in 2004. Good definitions and integration of the process was presented in a book by Hudson, Haas, and Uddin [40].

They differentiated physical assets from financial assets and called the new process *Infrastructure Management* (the book's title) in lieu of "Asset Management." They revised the book in 2013 under a more distinctive name, *Public Infrastructure Asset Management* [17].

Practical asset management has lagged because it is not possible to do "top-down" management without "bottom-up" data and information. Early efforts were undertaken by the Michigan DOT when a consultant prepared an impressive plan for asset management with beautiful input/output diagrams. A problem arose because there was no content in the individual boxes of the diagrams. As one can image, the effort never came to fruition. TRB efforts led by Dr. Sue McNeil, Mr. Steve Varnadoe, Lacey Love and others laid a good conceptual foundation, but the most definitive implementation has progressed stepwise for a working Maintenance Management System (MMS) or Pavement Management System (PMS).

42.1   Background

A major initiative in Europe by the Forum of European Highway Research Laboratories (FEHRL) has set out "Asset Management Challenges for Road Networks" [41]. Their "Vision for Asset Management" reads:

> In 2025 we will have a common understanding of an integrated and flexible approach towards asset management on a European level. Tools and flexible standards support an optimization of performances, risks, and cost of infrastructure within the modes, across the modes and between countries. Life-cycle management of asset systems and assets is a common practice. Asset management practices add a measurable value to different levels of economies, societies and environment. Best value is delivered to stakeholders.

Similar concepts of transportation asset management were presented in 2010 in Alaska, China, and Australia [39] as described below.

Transportation asset management systems (TAMS) have evolved over the past 20 years to the extent that many countries now have at least reasonably well developed component systems in place. These include pavement management systems (PMS), bridge management systems (BMS), traffic management systems (TMS), right-of-way features management systems (ROWMS), maintenance management systems (MMS), and others. Because pavements comprise a major part of the total road asset value,

PMS have developed faster and incorporated more technological advances. The component systems should function within a TAMS umbrella, but coordination and integration are not easy tasks.

While the structure of TAMS varies from agency to agency, examination of actual best practice internationally reveals that a comprehensive TAMS functions at three distinct but interrelated levels, all of which should exist within the agency's corporate business plan. The levels are:

- STRATEGIC level, where the business plan's mission statement, level of service and safety targets and policy objectives plus various economic, social, political, environmental and public or stakeholder group input factors are taken into account. Where long range financial forecasts and investment needs are carried out, and cost estimates are prepared to meet the defined targets. Current and future expected asset values should be included.
- NETWORK wide level, where alternative programs of asset preservation and network expansion are considered with performance estimates and life-cycle cost analyses (LCCA) are used to determine an optimal program for given budget(s) or funding levels.
- PROJECT level, where detailed physical and LCCA inputs are used to identify and implement the most economical, effective alternative for a project/link/site specific area.

While Project is an essential level in any comprehensive asset management system, it is not discussed in the following sections mainly because Part Four of this book is directed to the Project Level.

42.2 Framework for AMS

Figure 42.1 provides a framework for asset management: the main elements at each level are identified within boxes, and various selected or applied factors, models, constraints, forecasts, time horizons, etc. are listed at the right of the boxes. An integration platform is used as a mechanism for tying the road asset types, condition, etc. together by location plus asset value, level of service provided, and risk exposure, if possible. A brief description of the levels of TAMS main elements and various factors, models, etc. follows.

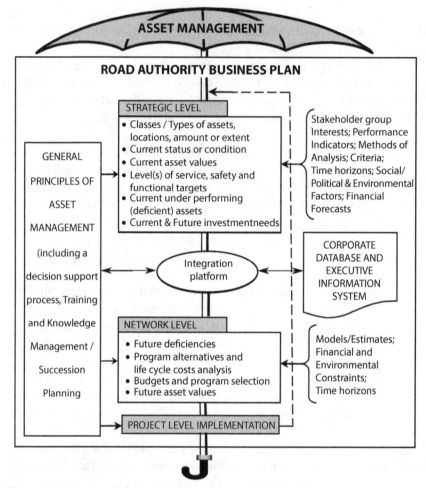

Figure 42.1 Framework for road asset management. After [39]

A key factor in both the road agency's business plan and in the TAMS itself is explicit recognition of stakeholder group interests and provision of service to them. Figure 42.2 illustrates provision of service as a central function and is directed to private and commercial road users. Providers of the service can range from a road agency to investor/concessionaires to managers for the road. Regulators, enforcement, standards, etc. are also associated with provision of service. Finally, preservation and efficiency requirements and measurable performance indicators are necessary to a properly functioning AMS.

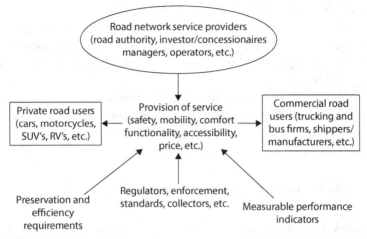

Figure 42.2 Stakeholder groups, service expectations and related factors in a TAMS. After [39]

42.3   Business Plan for AMS

Road agencies normally operate with a business plan in a business environment, which may be formally articulated (e.g., a mission statement followed by, for example, a 20 year vision of broad goals related to safety, environmental stewardship, mobility and accessibility, stakeholder group interests, etc.) or which may form an implicit operating environment of policies, standards, regulations, etc. The business plan or business environment should reflect the political, social, and economic responsibilities of people appointed or elected to act on behalf of the public.

42.4   General Principles of Asset Management Evolving from PMS

Figure 42.1 indicates that the general principles of asset management are applicable to all levels, a self-evident requirement, and that a decision support process plus training and knowledge management/succession planning functions should be included.

The decision support process should be based on the corporate data base and the executive information system derived from the data base. Essentially, decision support provides the necessary information, such as graphs, tables, forecasts, recommendations, etc., appropriate to the key elements identified in the strategic and network level of Figure 42.1.

For example, at the Strategic level, major decisions are likely meant to determine the tolerable shortfall between investment needs and financial forecasts over 10 to 20 years. At the network level, major decisions should involve approval of the works and associated programs based on the assigned budgets.

Aside from financial forecasts, a TAMS structure should be designed to provide necessary information for decision support. A comprehensive training component and a knowledge management/succession plan is also needed.

42.5   Early Positive Steps by DOTs

Most work toward the development of asset management (AMS) have begun with either pavement management or maintenance management as the foundation. To best illustrate the concept of moving toward asset management from a base of pavement management, a case study from the Idaho Transportation Department (ITD) provides an example. ITD first implemented integrated maintenance and pavement management systems in 2009–2010 [42-44]. ITD is one of several DOTs that have recognized the need to have a good PMS and a fully functioning MMS, and also to have good interaction between the two. This effort has been facilitated by IDT's implementing off-the-shelf PMS and MMS systems with a "System Foundation" from the same provider, all of whch are state-of-the-art and compatible data wise. The implementation team is fully integrated between IDT personnel and the software provider and is both fast and effective. Figure 42.3 illustrates the overlap and relationship between PMS and MMS. Although it duplicates Figure 31.1 in Part Six, it is also relevant to this chapter.

This process can be partially defined by six steps. The process then advances as the agency adds Safety, Bridges, Congestion, Signs, etc. Steps in the process are:

1. Establish or upgrade to a strong network level PMS using off-the-shelf software.
2. Define a common Linear Referencing System (LRS) that permits data exchange among systems, (PMS, MMS, and future add-ons such as BMS and Safety), including maintenance history, construction history, traffic database, etc.
3. Add implementation of an MMS integrated with the PMS using off-the-shelf software and the common LRS. (Steps 3 and 1 may be reversed if desired.)

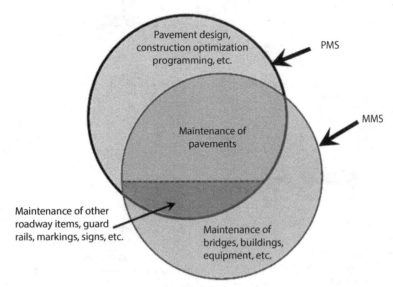

Figure 42.3 Interface of pavement management and maintenance management.
After [18]

4. Begin effectively using the PMS and MMS together. If you already have one of the systems in use as some agencies do, then move to add, integrate, and use the system. This may require a more complex interface with your old system, but it is well worth the effort.

5. Integrate the data on the common LRS for both PMS and MMS modules with appropriate transfer functions between the two systems. Share data to minimize double entry, provide efficiency, and minimize errors from creeping in between the various systems.

6. Provide the appropriate output results and reports to your DOT administrators and stakeholders for their use and proceed from there.

42.6   Maturing AMS

As the AMS process advances and reaches maturity, it will begin to look like the asset management Modular Framework shown in Figure 42.4. The large end piece of the puzzle, MMS, signified that this AMS started with core functions plus Maintenance Management, then followed Pavement Management, then Bridge Management. These, of course, can be added in

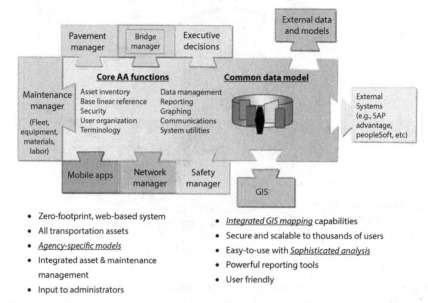

- Zero-footprint, web-based system
- All transportation assets
- *Agency-specific models*
- Integrated asset & maintenance management
- Input to administrators

- *Integrated GIS mapping* capabilities
- Secure and scalable to thousands of users
- Easy-to-use with *Sophisticated analysis*
- Powerful reporting tools
- User friendly

Figure 42.4 Asset Management Modular Framework. After [18]

any order. Each system or subsystem is shown as a plug-in module that can be modified, removed, or upgraded separately as desired or needed. Your AMS might start with core functions and data models plus any other systems in lieu of MMS, for example PMS and/or BMS or even the network manager.

42.7   Roadblocks to AMS Implementation

As with any developmental process, several internal activities in PMS/MMS expansion to AMS must be dealt with. Some of these involve research and development but did not get included on the FHWA PMS Roadmap [22] covered in Chapter 41, possibly because they extend beyond simple PMS, or possibly because there is not enough thinking outside the PMS box to see the true picture of DOT needs. Following are the main issues that must be dealt with.

1. Better techniques are needed to translate PMS information and predictions into maintenance, preservation, and rehabilitation actions.
2. Lack of detailed pavement maintenance data has long been a problem in PMS and limits optimal integration of maintenance and broader pavement budgets toward AMS.

3. LRS must be compatible across all modules or subsystems.
4. Models are needed that better predict the effect of maintenance and preservation actions on pavement performance. Some agencies predict pavement behavior based on design models that not only ignore maintenance but do not translate well to network level data.
5. Additional research is needed to define the effecs of various rehabilitation methods on pavement performance. These effects are often predicted without realistic data and/or good models of rehabilitation actions, and thus they create future errors.
6. Make data on major maintenance a part of structural history records based on a common LRS for the road and bridge network.
7. Using the factors and models outlined above, establish trade-offs among preservation, maintenance, and rehabilitation budgets.

42.8   Strategic Level

The incorporation of information on classes or types of assets, locations, amount or extent, and current status or condition is a mandatory requirement in any properly functioning AMS. Current asset value is desirable but not necessary. Levels of service, safety, and functional targets may be part of the business plan, or stated as policy objectives, but in any case, together with performance indicators and criteria (sometimes called "trigger values"), these provide the basis for identifying deficient or underperforming assets (e.g., length of pavement which exceed trigger values for smoothness and/or skid resistance; bridges which are functionally or structurally inadequate, etc.). Performance of assets over a stated time horizon can be estimated, and future investment needs to meet targets or policy objectives can be calculated. Using available financial forecasts, shortfall between needs and available funds can be identified.

42.9   Corporate Data Base and Executive Information System

The corporate/road authority's data base is a core part of any good TAMS. Normally the software packages for maintaining and using such data bases are acquired from vendors that specialize in the area. There are cases where a customized data base has been developed in-house, but the cost, time required, maintenance, and upgrades are not generally justified. A comparative assessment of the major internationally available TAMS vendor software packages is described in [17].

42.10   Network Level and Project Level

At the network level, the works program alternatives, the applicable time horizon, or program period are evaluated for life-cycle cost effectiveness within available budget(s). Environmental and other constraints also apply. Future deficiencies should be identified since available budgets are usually a constraint and future asset values should be estimated. The selected work in the program is then carried out at the project or site-specific level.

42.11   Summary

More detailed discussion of Asset Management is beyond the scope of this book, but to summarize, several state DOTs are already using PMS and MMS together to work toward a full AMS. As pointed out, roadblocks exist, but in most cases this combined approach is far ahead of and moving faster than generalized efforts to start an AMS from scratch or from the top down.

42.12   Websites Containing Transportation Asset Management Information as of 2014

AASHTO

Transportation Asset Management Today
Sponsored by the AASHTO Subcommittee on Transportation Asset Management
http://assetmanagement.transportation.org/tam/aashto.nsf/home

Transportation Asset Management in Australia,
Canada, England and New Zealand
International Scanning Tour (2005)

FHWA-PL-05-019
http://assetmanagement.transportation.org/tam/aashto.nsf/All+
    Documents/30F144B18E33667A8
52570A000468331/$FILE/TAM-International_Scan_Final_Report.pdf

Asset Management
http://www.fhwa.dot.gov/infrastructure/asstmgmt/

An Asset-Management Framework for the Interstate Highway System
NCHRP Report 632
http://onlinepubs.trb.org/onlinepubs/nchrp/nchrp_rpt_632.pdf

U.S. Domestic Scan Program: Best Practices in Transportation
Asset Management (NCHRP Project 20-68), 2007
http://onlinepubs.trb.org/onlinepubs/trbnet/acl/NCRHP2068_Domestic_
Scan_TAM_Final_Report.pdf

Association of Local Government Engineering New Zealand, Inc. and the
Institute of Public Works Engineering of Australia. 2006. *International*
Infrastructure Management Manual. Version 3.0, Association of Local
Government Engineering New Zealand, Inc.

World Bank. Road Asset Management
http://web.worldbank.org/WBSITE/EXTERNAL/COUNTRIES/
    SOUTHASIAEXT/EXTSARREGTOPT
RANSPORT/0,,contentMDK:20688593~menuPK:867153~pagePK:34004
    173~piPK:34003707~theSitePK:579598,00.html

References to Part Seven

1. Haas, R.C.G., W.R. Hudson, and J.P. Zaniewski, *Modern Pavement Management*, Krieger Publishing Company, Melbourne, Florida, 1994.
2. American Association of State Highway and Transportation Officials, "Mechanistic-Empirical Pavement Design Guide, A Manual of Practice," Interim Edition, July 2008.
3. Hudson, W.R., C. Monismith, C. Dougan, P Visser, "Using Pavement Management Data, To Calibrate and Validate the New MEPDG - An Eight State Study," FHWA Contract DTFH 61-05-C-00011, Federal Highway Administration, September 2006.
4. Pierce, L.M., K.A. Zimmerman, N. Saadatmand, "Use of Pavement Management Data for Calibration of the Mechanistic-Empirical Pavement Design Guide," 8th International Conference on Managing Pavement Assets, Santiago, Chile, 2011.
5. Hudson, W.R., C. Monismith, C. Dougan, P Visser, "Use of PMS Data for Performance Monitoring with Superpave as an Example," FHWA Contract DTFH61-98-C-00075, B98C75-007, Final Report, Volumes 1 and 2, FHWA, Washington, D.C., March 2002.
6. A. Bhargava, *et al.*, "Using an Integrated Asset Management System in North Carolina for Performance Management, Planning, and Decision Making," prepared for the 91st Transportation Research Board Annual Meeting, Washington, D.C., 2012.
7. Haas, Ralph and Afrooz Aryan, "Cross Optimization," Final Report to Ministry of Transportation Ontario, July 31, 2010.
8. Hudson, S.W., X. Chen, G. Cumberledge, and E. Perrone, "Pavement Performance Modeling Program for Pennsylvania," paper presented at the 74th Annual Meeting of the Transportation Research Board, Washington, D.C., January 1995.
9. Hudson, S.W., X. Chen, and E. Perrone, "Pavement Performance Modeling Program, User's Guide." Texas Research and Development Foundation, Draft Report No. 2., June 1994.
10. Prozzi, Jolanda, S. Grebenschikov, A. Banerjee, and J. Prozzi, "Impacts of Energy Developments on the Texas Transportation System Infrastructure," FHWA/TX-11/0-6513-1A, Center for Transportation Research at the University of Texas, Austin, 2011.
11. Smadi, Omar, "Knowledge based Expert System Pavement Management Optimization," paper for the 5th International Conference on Managing Pavements, Seattle, Washington, 2001.
12. Camba, J.C.M. and A.T. Visser, "Comparison of Pavement Management Outcomes from a Knowledge-Based Expert System with HMD-4," Paper for the 6th International Conference on Managing Pavements, Brisbane, Australia, 2004.

13. ESRI, "Highway Data Management in ArcGIS," an Esri' White Paper, Redlands, California, July 2010.

14. Andrén, Peter, "Development and Results of the Swedish Road Deflection Tester," Department of Mechanics, Royal Institute of Technology, Sweden, 2006.

15. Jitin, A., V. Tandon and S. Nazarian, "Continuous Deflection Testing of Highways at Traffic Speeds," Report FHWA/TX-06/0-4380-1, El Paso, Texas, October 2006.

16. Rada, Gonzalo, "The State-of-the-Technology of Moving Pavement Deflection Testing," FHWA Contract No. DTFH61-08-D-00025, Technical Report, Fugro Consultants, Inc. Austin, Texas, January 2011.

17. Uddin, W., Hudson, W.R., Haas, R., *Public Infrastructure Asset Management*, McGraw Hill Publishers, 2013.

18. Hudson, Stuart W., *et al.*, "Improving PMS by Simultaneous Integration of MMS," 8th International Conference on Managing Pavement Assets, Santiago, Chile, November 2011.

19. Better Roads, "Nanotechnology and the Future of Roads," June 2006.

20. Gopalakrishnan, K., B. Birgisson, P. Taylor, N.O. Attoh-Okine, *Nanotechnology in Civil Infrastructure, A Paradigm Shift*, Springer.com, 2011.

21. Steyn, Wynand JvdM, *Applications of Nanotechnology in Road Pavement Engineering*, Nanotechnology in Civil Infrastructure, Springer.com, 2011.

22. Zimmerman, K.A., L.M. Pierce, J. Krstulovich, "Pavement Management Roadmap, Executive Summary," FHWA Report # HIF-11-011, Applied Pavement Technologies, Urbana, Illinois, December 2010.

23. Mizusawa, D. and S. McNeil, "Generic/Methodology for Evaluating Net Benefit of Asset Management System Implementation," Journal of Infrastructure Systems, Vol. 15, 2009, pp 232–240.

24. Hudson, W.R., Hudson, S.W., Visser, W., and Anderson, V.L., "Measurable Benefits Obtained from Pavement Management," 5th Conference on Managing Pavements, Seattle, WA, 2001.

25. Akofio-Sowah, M.A. and A. Amekudzi, "Quantifying the Benefits of Managing Ancillary Transportation Assets," Transportation Research Board Annual Meeting, Washington, D.C., 2010.

26. AECOM, "Supplement to the AASHTO Transportation Asset Management Guide, Volume 2 – a Focus on Implementation," NCHRP, Transportation Research Board, Washington, D.C., 2010.

27. Cowe Falls, L.C., S. Khalil, W.R. Hudson, and R. Haas, "Long-Term Cost-Benefit Analysis of Pavement Management System Implementation," Third International Conference on Managing Pavements, Conference *Proceedings* 1, San Antonio, Texas, 1994, pp. 133–138.

28. FHWA/NHI Course #131105, "Analysis of PMS Data for Engineering Applications," 2006.

29. Anderson, V.L. and R.A McLean, "Design of Experiments, a Realistic Approach," Marcel Dekker, Inc., New York, 1974.

30. Paterson, William D.O, "Road Deterioration and Maintenance Effects," paragraph 2.1.2, World Bank, 1987.
31. Bartlett, Sue, Pinellas County, FL and Cox, Lydia, and Lorick, Harry, LA Consulting, Inc., Manhattan Beach, CA, AgileAssets User Group Meeting, Austin, Texas, 2007.
32. Haas Ralph, W.R. Hudson, L. Cowe Falls, "Evolution of and Future Challenges for Pavement Management," 8th International Conference on Managing Pavement Assets, Santiago, Chile, November 2011.
33. Haas, Ralph, "Evolution and Legacy of Pavement Management in Canada: a CGRA/RTAC/TAC Success Story," presented at the Innovative Developments in Sustainable Pavement Section, Annual Conference of the Transportation Association of Canada, Alberta, 2011.
34. Seedah, D., T. Owens, R. Harrison, "Evaluating Truck and Rail Movements along Competitive Multimodal Corridors," The University of Texas, Austin, Texas Department of Transportation, Report No. FHWA/TX-13/0-6692-1 published in TRR, Volume 2374, January 2014. pps 93–101.
35. Matthews, R., et al., "Estimating Texas Motor Vehicle Operating Costs: Final Report," TxDOT Study 05974-2, University of Texas, Austin Texas Department of Transportation, Federal Highway Administration, 2011, p 273.
36. Zaabar, Imen and Karin Chatti, "Calibration of HDM-4 Models for Estimating the Effect of Pavement Roughness on Fuel Consumption for U.S. Conditions," based on results from NCHRP Report 720, Transportation Research Record: Journal of the Transportation Research Board, Issue 2155, 2010, pp 116.
37. Fishburn, P.C., "Utility Theory for Decision Making," Research Analysis Corp, Final Report, PDF Url : AD0708563, McLean, Virginia, June 1970.
38. Stigler, George, "The Development of Utility Theory," Journal of Political Economy, Volume 58, #4, Columbia University, August 1950.
39. Haas, Ralph, "Transportation Asset Management Workshop," package prepared for Alaska DOT and PF, Anchorage, August 5, 2010.
40. Hudson, W.R., R. Haas, and W Uddin, Infrastructure Management, McGraw Hill Publishers, 1997.
41. Forum of European Highway Research Laboratories (FEHRL), "Asset Management Challenges for Road Networks: A Roadmap for Research," URL: http://www.fehrl.org/inde...ode=download&id_file=15017, Brussels, Belgium, 2013.
42. Perrone, Eric, "Bootcamp Training - Pavement Management," ITD MAPS Project, Version 1.0, conducted by AgileAssets, Inc for Idaho DOT, June 2010.
43. Scheinberg, Tonya and P. C. Anastasopoulos, "Pavement Preservation Programming: a Multi-Year Multiconstraint Optimization Methodology," Prepared for presentation at the 89th Transportation Research Board on TRR, January 2010.

44. Pilson, Charles, "Bootcamp Training Manual for Maintenance Management," ITD MAPS Project, Version 1.0, conducted by AgileAssets, Inc for Idaho DOT, updated January 2010.

Index

Also of Interest

Check out these other titles from Scrivener Publishing

Fracking, by Michael Holloway and Oliver Rudd, ISBN 9781118496329. This book explores the history, techniques, and materials used in the practice of induced hydraulic fracturing, one of today's hottest topics, for the production of natural gas, while examining the environmental and economic impact. *NOW AVAILABLE!*

Formation Testing: Pressure Transient and Formation Analysis, by Wilson C. Chin, Yanmin Zhou, Yongren Feng, Qiang Yu, and Lixin Zhao, ISBN 9781118831137. This is the only book available to the reservoir or petroleum engineer covering formation testing algorithms for wireline and LWD reservoir analysis that are developed for transient pressure, contamination modeling, permeability, and pore pressure prediction. *NOW AVAILABLE!*

Electromagnetic Well Logging, by Wilson C. Chin, ISBN 9781118831038. Mathematically rigorous, computationally fast, and easy to use, this new approach to electromagnetic well logging does not bear the limitations of existing methods and gives the reservoir engineer a new dimension to MWD/LWD interpretation and tool design. *NOW AVAILABLE!*

Desalination: Water From Water, by Jane Kucera, ISBN 9781118208526. This is the most comprehensive and up-to-date coverage of the "green" process of desalination in industrial and municipal applications, covering all of the processes and equipment necessary to design, operate, and troubleshoot desalination systems. *NOW AVAILABLE!*

Tidal Power: Harnessing Energy From Water Currents, by Victor Lyatkher, ISBN 978111720912. Offers a unique and highly technical approach to tidal power and how it can be harnessed efficiently and cost-effectively, with less impact on the environment than traditional power plants. *NOW AVAILABLE!*

Electrochemical Water Processing, by Ralph Zito, ISBN 9781118098714. Two of the most important issues facing society today and in the future will be the global water supply and energy production. This book addresses both of these important issues with the idea that non-usable water can be purified through the use of electrical energy, instead of chemical or mechanical methods. *NOW AVAILABLE!*

Biofuels Production, Edited by Vikash Babu, Ashish Thapliyal, and Girijesh Kumar Patel, ISBN 9781118634509. The most comprehensive and up-to-date treatment of all the possible aspects for biofuels production from biomass or waste material available. *NOW AVAILABLE!*

Biogas Production, Edited by Ackmez Mudhoo, ISBN 9781118062852. This volume covers the most cutting-edge pretreatment processes being used and studied today for the production of biogas during anaerobic digestion processes using different feedstocks, in the most efficient and economical methods possible. *NOW AVAILABLE!*

Bioremediation and Sustainability: Research and Applications, Edited by Romeela Mohee and Ackmez Mudhoo, ISBN 9781118062845. Bioremediation and Sustainability is an up-to-date and comprehensive treatment of research and applications for some of the most important low-cost, "green," emerging technologies in chemical and environmental engineering. *NOW AVAILABLE!*

Sustainable Energy Pricing, by Gary Zatzman, ISBN 9780470901632. In this controversial new volume, the author explores a new science of energy pricing and how it can be done in a way that is sustainable for the world's economy and environment. *NOW AVAILABLE!*

Green Chemistry and Environmental Remediation, Edited by Rashmi Sanghi and Vandana Singh, ISBN 9780470943083. Presents high quality research papers as well as in depth review articles on the new emerging green face of multidimensional environmental chemistry. *NOW AVAILABLE!*

Energy Storage: A New Approach, by Ralph Zito, ISBN 9780470625910. Exploring the potential of reversible concentrations cells, the author of this groundbreaking volume reveals new technologies to solve the global crisis of energy storage. *NOW AVAILABLE!*

Bioremediation of Petroleum and Petroleum Products, by James Speight and Karuna Arjoon, ISBN 9780470938492. With petroleum-related spills, explosions, and health issues in the headlines almost every day, the issue of remediation of petroleum and petroleum products is taking on increasing importance, for the survival of our environment, our planet, and our future. This book is the first of its kind to explore this difficult issue from an engineering and scientific point of view and offer solutions and reasonable courses of action. *NOW AVAILABLE!*